TACKLING TABLES

Using a Strategies Approach

Paul Swan

Tackling Tables: Using a Strategies Approach – 2nd Edition

First published 2007

Author: Paul Swan

ISBN 978-0-9585632-6-0

Printed by Success Print

The author wishes to thank Sharon Locke, Linda Marshall, Geoff White and Kellee Williams for their support and comments on drafts of the manuscript.

Copies of this book and others shown on page 4 may be purchased from good educational bookstores or by contacting the author at: pswan@iprimus.com.au

Structure of the Book

The ability to fluently recall the basic multiplication facts, or tables as they are widely known, is an important facet of mathematics. Drill and practice activities are often used to develop recall of these facts.

The purpose of this publication is to provide teachers with some alternatives to the standard drill and practice activities used in many classrooms.

This book is divided into several parts.

Section 1: Preparing to learn the basic facts. This section includes:

- Establishing what constitutes the basic facts,
- Examining the operation of multiplication,
- Reviewing the properties of multiplication, and
- Explaining strategies associated with the learning of basic multiplication facts.

In the **second section** includes:

- Approaches to teaching/learning the basic multiplication facts, and
- Ideas for assessing what students know and what they need to learn.

Many of the activities are designed to be used several times. To this end, blank copies of various pages have been provided so that teachers can rework and reuse the activities. Students may even use reproducible sheets to create new activities for other children in the class to complete. A small photocopy symbol appears on the top of each reproducible page.

The **third section** contains activities designed to build a bank of known facts in non-threatening situations.

The **final section** examines ways to increase the speed of response with basic multiplication facts. Activities are provided to assist students in developing fluency with basic number facts.

Some other books by the same author

I sometimes self publish and at other times I write for publishers. Here is a listing of my current titles.

Published by A–Z Type

Dice Dilemmas

Dice Dazzlers

Card Capers

Domino Deductions

Calculators

Turn & Learn

Published by RIC Publications (www.ricgroup.com.au)

Motivational Maths

Maths Investigations

Check Your Work

Number Grids

Maths Terms & Tables

MasterGrids

Patterns Upper

Patterns Middle

The hands on series. *Developing maths with …*

Unifix

Pattern Blocks

Base Ten

Contents

What are the Basic Multiplication Facts?

The basic multiplication facts are often referred to as the tables because they are often displayed in table format (See the layout to the right). Technically the basic multiplication facts refer to all single-digit multiplications, that is from 0 x 0 to 9 x 9.

These are depicted in the grid shown below. This formation is preferred over the traditional 'table' format.

x	0	1	2	3	4	5	6	7	8	9
0	0	0	0	0	0	0	0	0	0	0
1	0	1	2	3	4	5	6	7	8	9
2	0	2	4	6	8	10	12	14	16	18
3	0	3	6	9	12	15	18	21	24	27
4	0	4	8	12	16	20	24	28	32	36
5	0	5	10	15	20	25	30	35	40	45
6	0	6	12	18	24	30	36	42	48	54
7	0	7	14	21	28	35	42	49	56	63
8	0	8	16	24	32	40	48	56	64	72
9	0	9	18	27	36	45	54	63	72	81

Comparing Table Layouts

There are several reasons why this format is preferred. Setting the basic multiplication facts out in 'table format', as shown to the right, *hides the commutative property of multiplication*. Many students who learn the 'three times table' often cannot 'see' the relationship to other tables. For example, many students do not see that if they know 8 x 3, they should also know 3 x 8 and hence they have to learn twice as many multiplication facts.

The commutative property is an extremely important property of multiplication that states that the order in which a multiplication is performed does not affect the result. This property is discussed in detail later.

0 x 1 = 0	0 x 2 = 0	0 x 3 = 0
1 x 1 = 1	1 x 2 = 2	1 x 3 = 3
2 x 1 = 2	2 x 2 = 4	2 x 3 = 6
3 x 1 = 3	3 x 2 = 6	3 x 3 = 9
4 x 1 = 4	4 x 2 = 8	4 x 3 = 12
5 x 1 = 5	5 x 2 = 10	5 x 3 = 15
6 x 1 = 6	6 x 2 = 12	6 x 3 = 18
7 x 1 = 7	7 x 2 = 14	7 x 3 = 21
8 x 1 = 8	8 x 2 = 16	8 x 3 = 24
9 x 1 = 9	9 x 2 = 18	9 x 3 = 27
10 x 1 = 10	10 x 2 = 20	10 x 3 = 30
11 x 1 = 11	11 x 2 = 22	11 x 3 = 33
12 x 1 = 12	12 x 2 = 24	12 x 3 = 36

0 x 4 = 0	0 x 5 = 0	0 x 6 = 0
1 x 4 = 4	1 x 5 = 5	1 x 6 = 6
2 x 4 = 8	2 x 5 = 10	2 x 6 = 12
3 x 4 = 12	3 x 5 = 15	3 x 6 = 18
4 x 4 = 16	4 x 5 = 20	4 x 6 = 24
5 x 4 = 20	5 x 5 = 25	5 x 6 = 30
6 x 4 = 24	6 x 5 = 30	6 x 6 = 36
7 x 4 = 28	7 x 5 = 35	7 x 6 = 42
8 x 4 = 32	8 x 5 = 40	8 x 6 = 48
9 x 4 = 36	9 x 5 = 45	9 x 6 = 54
10 x 4 = 40	10 x 5 = 50	10 x 6 = 60
11 x 4 = 44	11 x 5 = 55	11 x 6 = 66
12 x 4 = 48	12 x 5 = 60	12 x 6 = 72

0 x 7 = 0	0 x 8 = 0	0 x 9 = 0
1 x 7 = 7	1 x 8 = 8	1 x 9 = 9
2 x 7 = 14	2 x 8 = 16	2 x 9 = 18
3 x 7 = 21	3 x 8 = 24	3 x 9 = 27
4 x 7 = 28	4 x 8 = 32	4 x 9 = 36
5 x 7 = 35	5 x 8 = 40	5 x 9 = 45
6 x 7 = 42	6 x 8 = 48	6 x 9 = 54
7 x 7 = 49	7 x 8 = 56	7 x 9 = 63
8 x 7 = 56	8 x 8 = 64	8 x 9 = 72
9 x 7 = 63	9 x 8 = 72	9 x 9 = 81
10 x 7 = 70	10 x 8 = 80	10 x 9 = 90
11 x 7 = 77	11 x 8 = 88	11 x 9 = 99
12 x 7 = 84	12 x 8 = 96	12 x 9 = 108

Some students who learn their 'tables' by chanting through a list of facts given in order often have to recite their way through part of the table list in order to arrive at the appropriate multiplication fact. For example, to find the result of multiplying six by eight these students recite the facts:

1 x 8 = 8
2 x 8 = 16
3 x 8 = 24
4 x 8 = 32
5 x 8 = 40

before reaching the desired 6 x 8 = 48.

Still another reason against presenting the multiplication facts in the typical 'table format' is that the ***patterns*** and ***relationships*** between basic multiplication facts such as the two, four and eight 'times tables' and the three, six and nine 'times tables' are less obvious.

The grid layout as shown below not only has the advantage of showing the commutative nature of multiplication but also it helps to link multiplication and division. In the example below 6 x 7 = 42 may be seen easily. What is equally important is that the factors of 42 are clearly depicted.

What about 10, 11 and 12?

Many books and charts go beyond single-digit multiplication and include facts up to 12 x 12. It makes sense to consider 'multiplying by ten' as the ability to multiply and divide by ten and powers of ten such as 100 and 1000. Even though the 'ten times table' is not part of the basic multiplication facts grid, at times reference will be made to the 'ten times table' in this book as the patterns are interesting and knowing the 'ten times table' can assist in learning the 'five and nine times tables'.

The 'eleven times table' forms a pattern and may be worth studying from that standpoint but there seems little need to know the eleven facts off by heart as the pattern makes it easy to generate the result.

The ability to quickly multiply by twelve was essential when imperial measurement system was used and before the introduction of decimal currency. The 'twelve times table' has less application today.

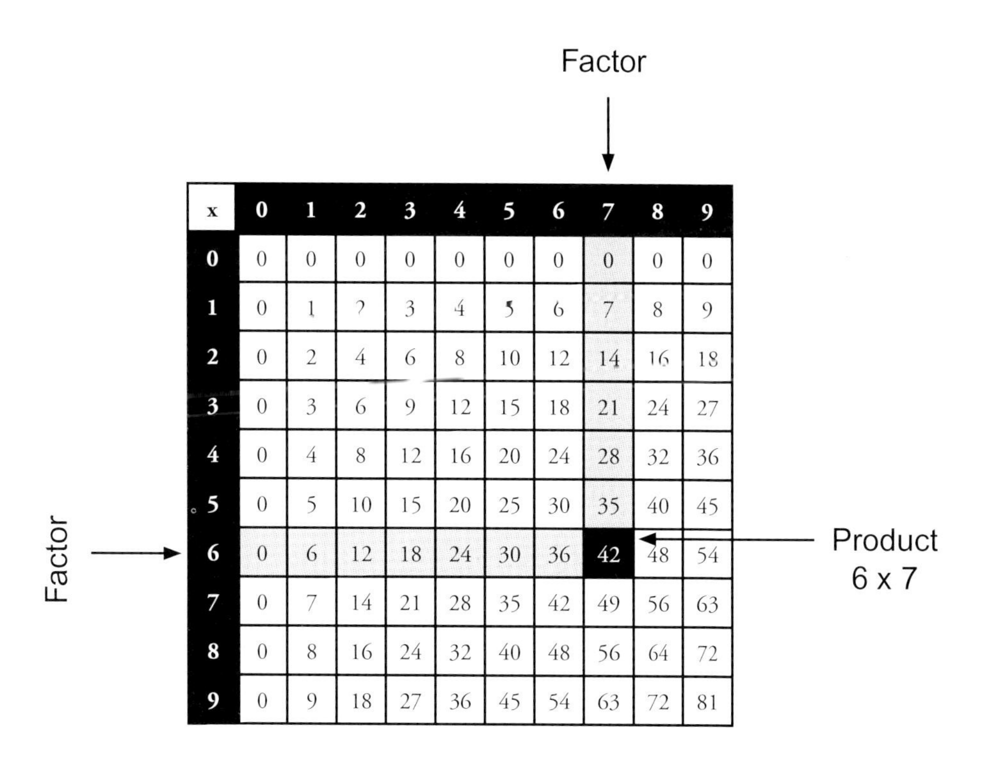

x	0	1	2	3	4	5	6	7	8	9
0	0	0	0	0	0	0	0	0	0	0
1	0	1	2	3	4	5	6	7	8	9
2	0	2	4	6	8	10	12	14	16	18
3	0	3	6	9	12	15	18	21	24	27
4	0	4	8	12	16	20	24	28	32	36
5	0	5	10	15	20	25	30	35	40	45
6	0	6	12	18	24	30	36	42	48	54
7	0	7	14	21	28	35	42	49	56	63
8	0	8	16	24	32	40	48	56	64	72
9	0	9	18	27	36	45	54	63	72	81

Before the Basic Multiplication Facts

Most of the difficulties that students experience when learning the tables may be traced back to inadequate understanding of the concept of multiplication and the associated multiplication properties. What follows is a brief overview of the concept of multiplication and the properties of multiplication. It is by no means comprehensive or exhaustive. For a detailed explanation consult a text on teaching primary mathematics, such as:

Booker, G., Bond, D., Sparrow, L., & Swan, P. (2004). *Teaching primary mathematics* (third edition). NSW: Pearson.

The Multiplication Concept

There are several underlying ideas that help form the multiplication concept. These spring from situations and problems that students will encounter both in and out of school.

Repeat Equal Quantities

Students who are additive thinkers often view multiplication simply as repeated addition. While this represents one view of multiplication, it is extremely narrow, and unless the concept of multiplication is developed further, students will struggle learning the basic multiplication facts. The phrase 'repeated addition' does not really describe what is involved in multiplying, as you are really adding or repeating equal quantities. For example, adding four, six times may be thought of as 4 + 4 + 4 + 4 + 4 + 4 or six sets of, or groups of, four.

Skip counting and using the constant feature on the calculator relate to this aspect of multiplication. A calculator may be set to repeatedly add equal quantities by using the inbuilt constant feature. For example, most calculators may be set to count by four using the following keystrokes:

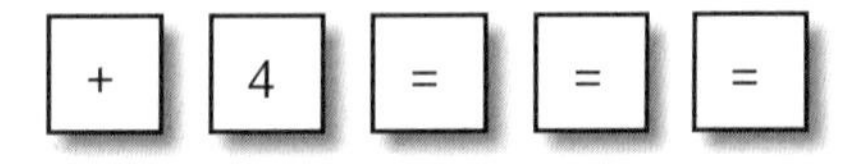

In simple terms, when the number and size of groups are given then multiplication is implied. There are subtle differences between situations that lead to this type of multiplication. Likewise subtle differences in situations or the way problems are presented can make a difference. Consider the following problems:

"If five children have three marbles each, how many marbles are there altogether?" and

"If there are three marbles per child, how many marbles would five children have altogether? "

The first implies the multiplication based on the idea of equal quantities being repeated. The second however involves a different idea of multiplication – rates.

Rates

Rates problems typically arise in 'real world' situations, such as buying fruit and vegetables. For example, calculating the price of 2·3 kg of apples at $2·99 per kilogram involves the use of multiplication. Calculating the cost of buying fuel where the price per litre and the number of litres is provided is another example.

Array

An understanding of the array or 'row by column' form of multiplication is extremely important when it comes to learning the multiplication tables. **The use of an array model is the best way to develop an understanding of the commutative property, that is, the order in which the multiplication occurs does not change the result. The commutative property in turn may be used to reduce the number of individual facts to be learned.**

Typical situations that lead to this form of multiplication include problems that involve rows and columns. For example, "A muffin tray holds four rows of three muffins. How many muffins in all?"

The muffin tray would look like this.

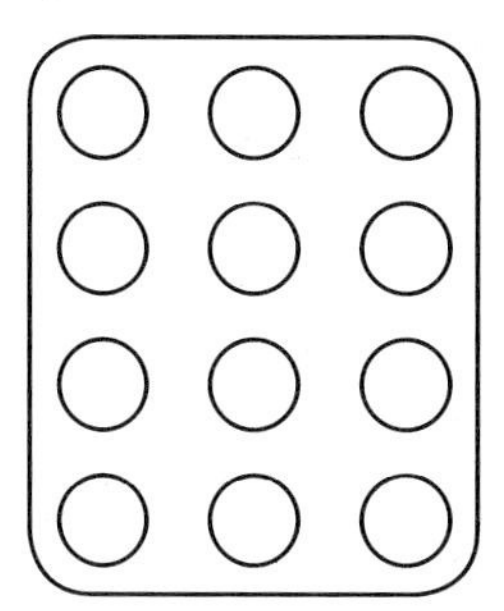

The number sentence that would accompany this situation would be 4 x 3 = 12.

Alternatively if the muffin tray was rotated 90 degrees we would speak of three rows of four. The number sentence would read 3 x 4 = 12.

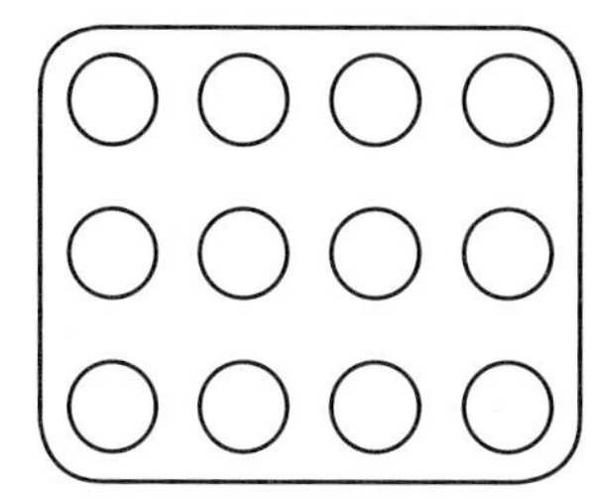

Obvious connections may be made to calculating the area of rectangular regions.

Combinations

As the name suggests, combination-type problems involve situations where there are two or more sets of objects and one object from one set is combined with another and so on. The aim is to calculate the total number of combinations. Consider the following situation. A canteen offers three kinds of lunch: a chicken wrap, salad roll or vegie burger and two kinds of drink: a bottled water or a juice. How many different lunch orders can be made from the selection that is offered? The situation may be illustrated using a tree diagram [See diagram top right].

The number sentence that would relate to this situation is 3 x 2 = 6.

Tree diagram showing various lunch combinations.

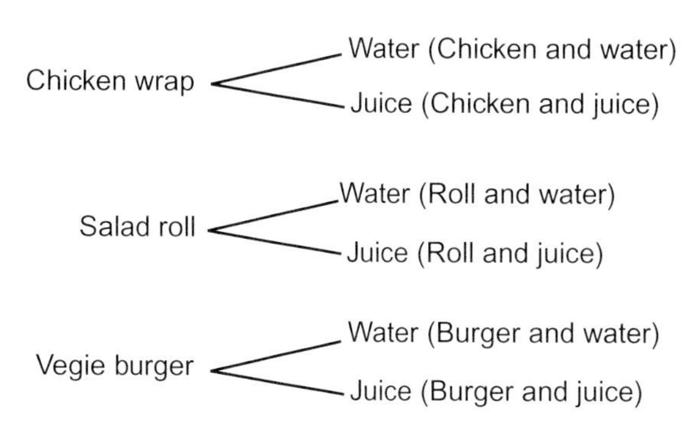

The key point is that students are exposed to a variety of multiplication situations so they build a strong understanding of the multiplication concept. To illustrate, consider how skip counting is related to the concept of repeating equal quantities.

To multiply 3 by 4 using a skip counting approach would involve counting by 4 three times, saying: "4, 8, 12." This is based on the idea of counting three groups of four. This approach may be supported via the use of technology where a calculator is set to constantly add 4.

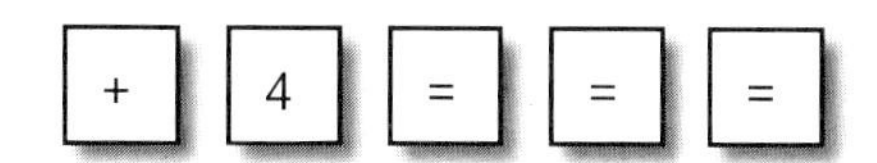

To multiply 3 by 4 a student applying array based thinking may picture three rows of four.

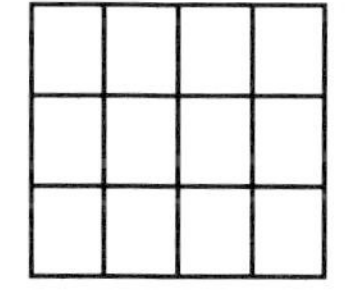

Properties of Multiplication

Developing a good understanding of the properties of multiplication will assist students to learn the basic multiplication facts. A brief overview of each of these properties follows.

The Multiplication Property of Zero

This is one of the simplest properties of multiplication. Essentially **any number multiplied by zero is zero.** Some students confuse the multiplication property of zero with adding zero. Adding zero to a number does not change the number, that is 7 + 0 = 7, whereas multiplying a number by zero produces a result of zero. Students who struggle with learning the basic multiplication facts often try to remember individual zero facts such as 3 x 0 = 0 and 7 x 0 = 0, rather than simply remembering that **anything multiplied by zero is zero. The property is summarised as $a \times 0 = 0$, or $0 \times a = 0$.**

Students who experience difficulty understanding this property can be exposed to real world situations that highlight the fact that the answer is zero. For example, "Jeremy put zero lollies into each of six bags. How many lollies did he use?" While students may find this sort of problem unusual or humorous it will help highlight the multiplication property of zero.

Similar examples related to the array model of multiplication may be used. For example, zero rows of any size will equal zero.

Division by Zero

It is important that the relationship between multiplication and division be emphasised when learning the basic multiplication facts. However division by zero does not make sense. Using repeated subtraction of equal quantities doesn't make sense. For example, in the question 8 ÷ 0, no matter how many times zero is subtracted from 8 you will never reach zero. **Division by zero is described as being undefined.**

The Multiplication Property of One

Any number multiplied by one is itself. Instead of having to learn a variety of 'multiplication by one' facts all students need to do is make use of the multiplication property of one. A good grasp of this property will assist in later years when working with equivalent fractions as essentially converting from one fraction to an equivalent fraction involves multiplying by one in another form such as $^3/_3$ or $^5/_5$. **The property is summarised as $a \times 1 = a$.**

The Associative Property of Multiplication

This property comes into play when three or more numbers are multiplied. As such it is not directly applicable to the learning of the basic multiplication facts, but ***it is an extremely useful property that underpins mental strategies associated with multiplication.*** For example, when multiplying 3 x 2 x 5, it is probably easier to calculate the 2 x 5 part first as this produces a 'nice number', 10. Multiplying 3 x 10 is fairly simple if the pattern of multiplying a number by ten has been developed. This property is summarised as $(a \times b) \times c = a \times (b \times c)$.

The Distributive Property of Multiplication Over Addition

A calculation may be broken apart in order to make it easier to complete. For example, a student who is building up the multiplication fact 8 x 7, may partition or break the 7 into 5 and 2 and then complete the calculation in the following manner.

$$\begin{aligned} 8 \times 7 &= 8 \times (5 + 2) \\ &= (8 \times 5) + (8 \times 2) \\ &= 40 + 16 \\ &= 56 \end{aligned}$$

This property is particularly useful when applying the mental strategy of relating an unknown fact to a known fact in order to work out the answer.

Cutting up a rectangular model helps to illustrate this property.

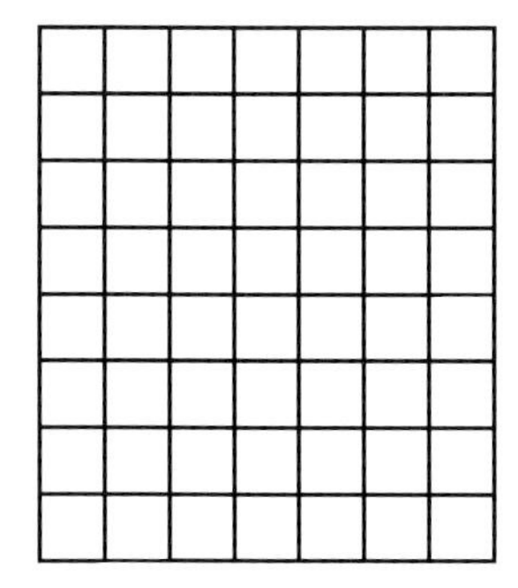

8 x 7

This rectangular region may be split into two parts.

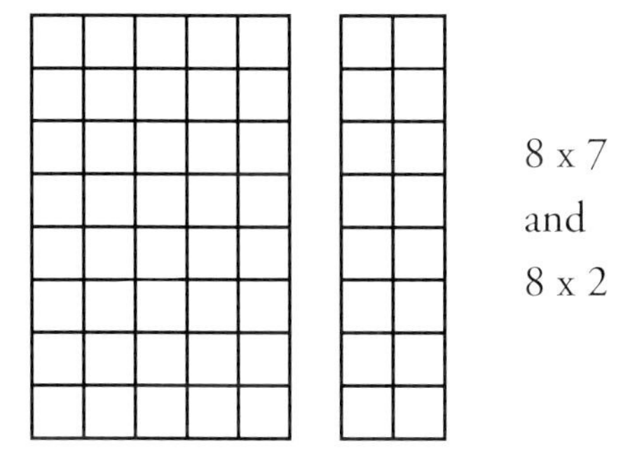

This property is summarised as:

$$a \times (b + c) = (a \times b) + (a \times c)$$

An understanding of the distributive property can help students learn the more difficult basic multiplication facts.

The distributive property is extremely important as it also underpins the development of the standard multiplication algorithm. Later the distributive property is used in algebra.

The Commutative Property of Multiplication

The commutative property of multiplication states that **numbers may be multiplied in any order without changing the result.** An understanding of this property is extremely helpful in learning the basic multiplication facts as it means that students have less facts they need to remember. *Once students learn one fact such as 7 x 3 = 21, they really should know the related fact, 3 x 7 = 21.* The commutative property is what underpins this.

It is well worth spending time establishing this property before starting to learn the basic multiplication facts. ***This property may be developed from the array model of multiplication.***

Students may be encouraged to stick dots or draw objects onto card to illustrate a multiplication sentence. For example, the following diagram illustrates the multiplication sentence 4 x 6, that is 4 rows of 6.

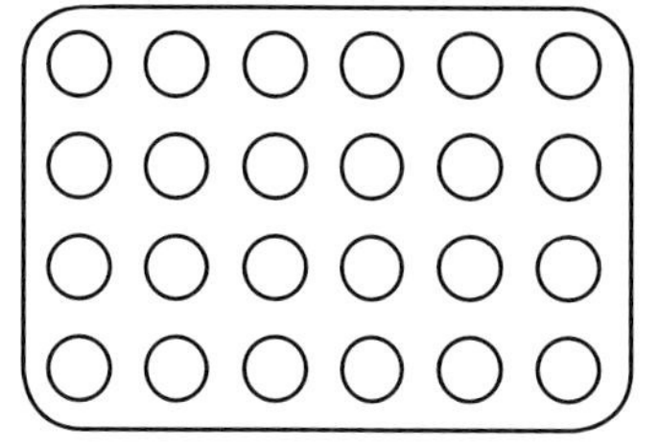

Rotating the card 90° illustrates the multiplication sentence 6 x 4, that is 6 rows of 4.

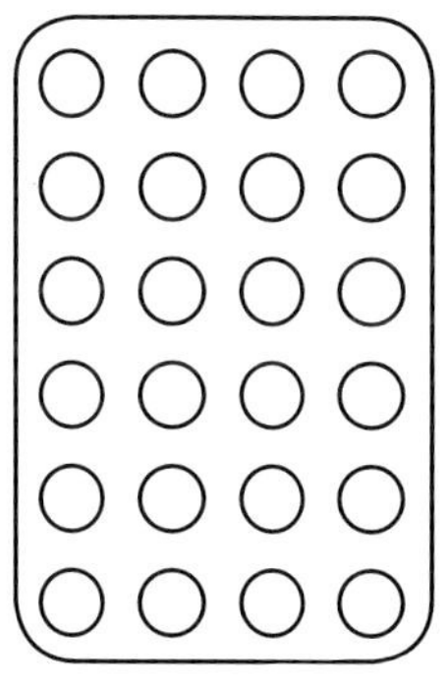

Encourage students to talk about the different situations that are illustrated. Further activities involving cutting out grid or graph paper and working with pegboards and geoboards will help illustrate the commutative property. These may be used to form arrays and may be easily be rotated. Everyday items such as muffin trays and egg cartons may be used to develop understanding of the commutative property.

Note: While multiplication may be thought of being commutative when simply considering number sentences, it may not necessarily be the case in particular situations. The number sentences may be describing different situations. For example, taking three tablets six times a day is certainly different from taking six tablets three times a day. While in both cases you will have taken 18 tablets, you may get sick, or worse die, if the order is changed!

The above properties may be combined to reduce the number of multiplication facts that need to be learned.

Reducing the Memory Load

The:

- multiplication property of zero
- multiplication property of one, and
- the commutative property

may be combined to reduce the number of individual basic multiplication facts to be learned.

Consider the following grid.

x	0	1	2	3	4	5	6	7	8	9
0	0	0	0	0	0	0	0	0	0	0
1	0	1	2	3	4	5	6	7	8	9
2	0	2	4	6	8	10	12	14	16	18
3	0	3	6	9	12	15	18	21	24	27
4	0	4	8	12	16	20	24	28	32	36
5	0	5	10	15	20	25	30	35	40	45
6	0	6	12	18	24	30	36	42	48	54
7	0	7	14	21	28	35	42	49	56	63
8	0	8	16	24	32	40	48	56	64	72
9	0	9	18	27	36	45	54	63	72	81

Multiplication property of one

Multiplication property of zero

Understanding the ***multiplication property of zero*** reduces the 100 table facts that need to learned by 19. Only 81 left to learn. Applying the ***multiplication property of one*** reduces the the number of multiplication facts to be learned by a further 17 facts, leaving 64 to learn.

Applying the ***commutative property of multiplication*** reduces the sixty-four multiplication facts to be learned by almost half. The black squares with white numbers represent the facts that are duplicated. The black numbers in white boxes are the ones that need to be learned.

The white numbers in black boxes represent the individual facts to be learned.

x	0	1	2	3	4	5	6	7	8	9
0	0	0	0	0	0	0	0	0	0	0
1	0	1	2	3	4	5	6	7	8	9
2	0	2	4	6	8	10	12	14	16	18
3	0	3	6	9	12	15	18	21	24	27
4	0	4	8	12	16	20	24	28	32	36
5	0	5	10	15	20	25	30	35	40	45
6	0	6	12	18	24	30	36	42	48	54
7	0	7	14	21	28	35	42	49	56	63
8	0	8	16	24	32	40	48	56	64	72
9	0	9	18	27	36	45	54	63	72	81

Notice that many of the tables to be learned are:

- the square numbers, (see dark grey squares in the chart below.)
- multiples of two
- the multiples of five and
- the multiples of three and nine.

x	0	1	2	3	4	5	6	7	8	9
0	0	0	0	0	0	0	0	0	0	0
1	0	1	2	3	4	5	6	7	8	9
2	0	2	4	6	8	10	12	14	16	18
3	0	3	6	9	12	15	18	21	24	27
4	0	4	8	12	16	20	24	28	32	36
5	0	5	10	15	20	25	30	35	40	45
6	0	6	12	18	24	30	36	42	48	54
7	0	7	14	21	28	35	42	49	56	63
8	0	8	16	24	32	40	48	56	64	72
9	0	9	18	27	36	45	54	63	72	81

There are various patterns and thinking strategies that may be used to help students learn these basic multiplication facts.

Strategies That Assist in Learning the Basic Multiplication Facts

A variety of strategies, some of which are based on the properties of multiplication discussed earlier, will assist students trying to learn the basic multiplication facts. ***Most important among these strategies is the ability to double and halve.***

Doubling

Doubling may be used to develop the 'two times table'. For example, 6 x 2 or 2 x 6 may be thought of as double six. Many students seem to have a natural ability to double. Some of this stems from the common practice of counting in twos and some of the chanting that takes place in the playground. For example, "two, four, six, eight who do you appreciate?" Time spent developing students' ability to double will be well spent, as *doubling is the strategy that is the key behind the cluster approach to learning the basic multiplication facts.*

The doubling strategy may be applied to the development of the 'four times table'. For example, to work out 7 x 4 or 4 x 7 all a student would need to do is double seven and then double again. That is, 7, 14, 28, so 4 x 7 is 28.

Essentially if you can *double and then double again*, you can 'work out' the 'four times table'. In order to apply this strategy requires an understanding of the commutative property. In the example above rather than think of seven fours a student would need to 'turn around' the calculation so that the focus is on the seven. Essentially the calculation becomes 4 x 7, which then changes to 2 x 2 x 7 and the via the use of the associative property becomes 2 x 7 (which is 14) and then the fourteen is doubled (or multiplied by two). In this case the doubling strategy is teamed with two properties of multiplication in order to 'work out' the 'four times table.'

Doubling, doubling and doubling again is a strategy that may be used to 'work out' the 'eight times table'. For example to 'work out' 8 x 7 a student applying the double, double, double strategy would need to think 7 x 8 may be turned around to 8 x 7. Starting with 7 the student would double to produce 14, then double again to make 28 and then double again to arrive at the answer of 56.

While this strategy is extremely powerful for 'working out' the 'two, four and eight times table', the aim should be for students to develop fluency with the basic multiplication facts. The issue of fluency and how to develop it is discussed later. ***The focus of this section is on developing a bank of facts.*** Suggestions for increasing the speed of recall will be provided later.

Use of Patterns

Strictly speaking the 'ten times table' is not part of the basic multiplication facts as these are restricted to single-digit by single-digit multiplication. However, it is not unreasonable to suggest that students learn to multiply by ten and later by powers of ten such as 100, 1000 and 0·1. A calculator is an ideal tool for generating patterns. A basic calculator may be set to constantly multiply by ten using a keystroke sequence similar to the one shown.

Students observing this pattern may describe it as 'adding a zero'. Discourage this thinking. It may appear to work with whole numbers but this 'short-cut method' soon breaks down when dealing with decimal numbers such as when multiplying 0·3 x 10. Encourage the students to think that when multiplying by ten all the digits move one place to the left. Similarly multiplying a number by 100 moves the digits two places to the left. Knowing the 'ten times table' opens the way for developing the 'five times table'. A halving strategy may be linked to the 'ten times table' in order to produce the 'five times table.' For example, to work out 8 x 5 a student may think of half of 8 x 10.

Another approach would be to make use of the pattern that exists within the 'five times table'. The 5, 0, 5, 0, 5, 0 pattern produced when counting in fives will assist students to learn the 'five times table'. Colouring in the multiples of five on a number grid will help some students 'see' the pattern. Links to the clock may be made also.

Digit patterns exist within the 'three and nine times tables'. While these patterns may not directly help students learn the 'three and nine times tables' they do provide a vehicle for exposing students to these basic multiplication facts. These patterns will be examined later in the activities section.

The 'nine times table' may also be *connected* to the 'ten times table'. ***Once a basic multiplication fact is known, it may be used to work out another fact.*** For example, to work out 6 x 9 a student may think of 6 x 10, which is 60 and then subtract 6 in order to arrive at an answer of 54. Notice how one fact may be used to derive another. This leads to another powerful mental strategy - relate to a known fact.

Relate to a known fact

This is an extremely powerful strategy. However, there is a catch associated with its development. ***In order to make use of this strategy the student must 'know' some basic multiplication facts.*** Hence this strategy has been left to this point, after students have learned some basic facts.

Askew (1998) illustrated the use of this strategy and how it may be used to learn new basic multiplication facts using the following diagram.

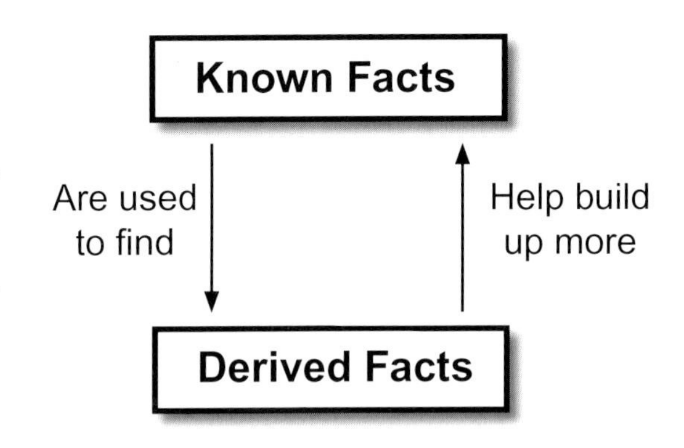

Askew, M. (1998). *Teaching primary mathematics: A guide for newly qualified and student teachers.* London: Hodder & Stoughton.

As the diagram illustrates, students may draw on their bank of known facts in order to 'work out' or derive new facts. For example, if a student does not 'know' 6 x 8 he/she can use a known fact, 5 x 8 as the basis for 'working out' 6 x 8. The thinking that would accompany the use of this strategy might run along the following lines; 'Five eights are 40 and another 8 is 48.'

Clearly the more facts that a student knows the larger the selection that may be used as the basis for deriving new facts. ***The use of this strategy should be seen as a 'stepping stone' to developing fluency with the basic multiplication facts.*** For example, knowing the square numbers, that is 1 x 1, 2 x 2, 3 x 3, 4 x 4 ... provides a base from which many basic multiplication facts can be derived. It is fairly simple to 'work out' 6 x 7 if you know 6 x 6 is 36 and then add another 6 to make 42. The fact that 6 x 6 = 36 has a language pattern, "six sixes are thirty-six" helps you to remember it. When drawn up into an array students can see that square numbers are 'square', hence the name.

Putting it all together

Combining the properties of multiplication and various mental strategies will help students to learn the multiplication facts. Eventually, however, the aim is for students to become fluent in using the basic multiplication facts.

Developing Fluency with the Facts

The aim of learning any basic fact should be to develop fluency. **The term 'fluency' describes the ability to access the required fact quickly without having to use a strategy to re-create it.** The term fluency has been chosen rather than 'instant' or 'automatic recall' because it relates to being fluent with language. When speaking or reading, sometimes a word is on the tip of your tongue, or you hesitate slightly, or substitute a word with a similar meaning. However, the flow of words and the meaning of what is said is not lost. The case is similar when utilising the basic multiplication facts. One of the few times a person is expected to recall 'the tables' out of context is in the classroom. All other uses of 'tables' are related to contexts. For example, in game situations such as playing darts, you may be required to work out triple 9 (clearly a basic multiplication fact) and combine it with 17 and 12.

Instant recall?

While recall is required, I have deliberately chosen not to add the words 'instant' or 'automatic' in front of it. To do so would imply a rate of speed that inhibits rather than assist students to recall their basic multiplication facts. Typically, research studies which make use of the terms 'instant recall' and 'automatic recall' allow students three seconds to respond to a basic facts question.

When students are put under the pressure of timed tests, rather than look for relationships and connections in number, they adopt the first method that springs to mind - no matter how inefficient. Further, the pressure of timed tests puts students under stress and they often panic, which clouds their ability to think. Often when you miss an early question in a timed test there is a cascade effect and you then miss most of the subsequent questions. Clearly there is a need to develop fluency in recalling the basic multiplication facts, but *timed tests often hinder rather than support the development of fluency.*

Rote learning?

Another phrase that is often associated with learning basic number facts is rote learning. Throughout this publication, the term memorisation is used as it better describes the intent of the activities contained in this book.

McIntosh (2005) described memorisation as:

> The committing to memory by students the relationships and number facts, which, if forgotten, they could efficiently recreate. It therefore assumes understanding. It is the opposite of rote-learning which is the committing to memory by students of relationships and number facts which, if forgotten, they could not efficiently re-create. It accompanies lack of understanding. (p. 5)

McIntosh, A. (2005). *Developing computation.* Department of Education Tasmania. Hobart Tasmania.

The key idea is that ***memorisation assumes understanding.*** A student who understands the connections and relationships between the various basic multiplication facts will soon be able to reconstruct the 'forgotten fact' efficiently using a combination of strategies and multiplication properties.

In the head, with the head

Beishuizen (1997) summarises the learning of any basic number facts nicely by separating out the idea of mental mathematics 'in the head' (the recall of facts) and mental mathematics 'with the head' (the use of strategies for calculating). Clearly both are required if a students is to efficiently work with and extend the basic multiplication facts.

Beishuizen, M. (1997) Mental arithmetic: Mental recall or mental strategies. *Mathematics Teaching 160*, 16-19.

Approaches to Learning the Basic Multiplication Facts

There are several different techniques that are used for teaching and learning the basic multiplication facts. *All are based on students having gained understanding of the properties of multiplication and the strategies described earlier in this publication.* There are many principles and common elements found in each of these approaches that may be adopted to design an approach that suits your students. Some of the approaches are more suited to older students who for one reason or another have not developed fluency with the basic multiplication facts. *Other approaches are more suited to the initial learning of the basic multiplication facts and, as such, would require that* ***the whole school adopt the same method, or at least the common elements.***

The cluster approach

The cluster approach is based on learning clusters or groups of facts rather than learning the 'two times table', then the 'three times table', the 'four times' and so on. ***The justification for adopting this method is that learning the tables facts in order does not assist students to spot patterns within the tables.*** Often when learning the multiplication facts in order (from the 'two times' to the 'ten times') the basic multiplication facts are presented in 'table form', effectively disguising the commutative property of multiplication and any supportive patterns found by clustering the 'table facts'.

Even within the cluster approach different clusters may be used. Here are some variations.

Clustering 2, 4, 8 and 3, 6, 9

The 'two, four and eight times tables' may be learned as one cluster and the 'three, six and nine times tables' are learned as another cluster. ***In order to use this approach students need to be able to 'double' and then double again.*** For example, if a student can double then all the 'two times tables' may be generated. To work out 2 x 8, all you need to do is double 8. To work out 7 x 2 just double 7. Two pieces of knowledge are required to reach this point. The first is an ability to double. This generally comes from experience with counting by two. ***The second idea that students need to grasp is the commutative property of multiplication.*** That is multiplying 7 by 2 gives the same result as multiplying 2 by 7.

- If you can double that means you can work out your 'two times table'.
- If you can double and double again then you can work out your 'four times table'.
- If you can double and double and double again then you can work out your 'eight times table'.

Similar logic may be applied to the three and the 'six times table'. Once you have learned the 'three times table', if you can double then you can work out the 'six times table'.

Co-operation between teachers of different year levels is required if this approach is adopted because it means the traditional approach, beginning with the 'two times table' and working in order through to the higher 'tables' is abandoned.

Clustering up to 5 x 5 and beyond

The approach recommended in the *First Steps in Mathematics* project involves beginning with the twos, fours and fives and then the threes, but only as far as 5 x 5. The facts are then compiled into a grid. Through the application of the commutative property and various activities such as skip counting, fluency with the basic multiplication facts to 5 x 5 is established. Through the use of patterns, properties of multiplication and mental strategies, the rest of the 'tables' are developed.

For a comprehensive summary of this approach see pages 190 - 191 of:

Department of Education and Training of Western Australia. *First Steps in Mathematics: Number Understand Operations, Calculate and Reason about Number Patterns.* Melbourne: Rigby.

Summary

To use the cluster approach you need to:

- **be able to double**
- **know how the commutative property of multiplication works**
- **know some facts to start with, eg the three times table.**

McIntosh Teaching Sequence

As stated previously, few authors advocate learning the 'tables' in order, progressing from 0 – 9 (or 12) in order. McIntosh is no exception. He outlined a teaching sequence for learning the basic multiplication facts that began with the two and three times table, then the 'one and zero times tables'. The entire sequence may be found on page 9 of McIntosh, A. (2004). *Mental computation: A strategies approach: Module 3 basic facts multiplication and division.* Hobart: Department of Education and Training.

The McIntosh sequence splits all table facts into two halves – the first five multiples and then the second five multiples. Teachers may then break the task of learning the table facts into smaller more manageable chunks.

In order to apply the McIntosh approach, agreement would need to be reached at a school level so that the transition between one year and the next is smooth.

Alternative Approaches

Some teachers believe that after coming to an understanding of the multiplication property of zero and one students should learn the 'square numbers' and then use these as the basis for deriving other basic multiplication facts. This method makes use of the 'relate to a known fact' strategy outlined earlier.

Catering For Individual Students

In every class, students will have:

- **a different understanding of the multiplication properties,**
- **a different range of strategies and**
- **different levels of fluency with the basic multiplication facts.**

Rather than just embark on teaching the 'six and seven times tables' because that is what is expected that year, why not find out what the students can do? Otherwise some, who already know the 'six and seven times table', will be thoroughly bored and others who struggle with the 'three times table' will be thoroughly out of their depth.

A 'table map' [See Table Map Activity, p. 21] can be created for each student so that both the teacher and the student are aware of strengths and weaknesses. ***While this may appear time consuming, it is well worth taking the time to complete this exercise. It will give you a picture of how individual students are faring and how the whole class is progressing. This will help develop a plan suited to the needs of individuals and groups within the class.***

Experience suggests that in most classes groups of students will have similar difficulties so you can group them together to work on those areas causing concern. For example, in most classes I work with, 1 in 3 students do not understand the commutative property. That represents ten students in every class of thirty! These students have not grasped the concept behind the commutative property and are therefore doing twice as much work as required. It would make sense to group these students together in order to help them understand the commutative property and thus overcome one roadblock to fluency.

Regularly I find 'weaker' students endeavouring to 'remember' individual table facts such as 1 x 3, 1 x 5, 1 x 6, 1 x 7 instead of applying the multiplication property of one. In effect these 'weaker' students are doing more work than those students who have learned to look for patterns, made use of properties of multiplication and developed mental strategies for multiplying.

All students, especially those who have not managed to learn their basic multiplication facts using other approaches, will benefit from making connections between interrelated facts – or fact families.

A Fact Family Approach

Students can be taught to learn families of facts rather than a set of unrelated facts. For example, when learning 4 x 3, it would make sense to relate this fact with 3 x 4 (emphasising the commutative property) and the associated division facts; 12 ÷ 4 = 3 and 12 ÷ 3 = 4, as these emphasise the relationship between multiplication and division. Depending on the nature of the learner, it would make sense to relate the division family to fractions by adding the related facts $^1/_4$ of 12 is 3 and $^1/_3$ of 12 is 4. In this way, students may begin to see connections between fractions and division.

Askew and Ebbutt (n.d) refer to the learning of 'triples' and suggest that students should be able to link three numbers that are written, for example, 3, 8 and 24. They extend this idea to what they refer to as "free gifts".

> Know one, get one free! Free gifts are the additional calculations that come 'free' when you know one number fact. This strategy depends on knowledge of the relationships between operations. So, if I know that 7 x 8 is 56, then I have the three free gifts: the answers to 8 x 7, 56 ÷ 7 and 56 ÷ 8. This is far easier than committing 'seven eights' to memory in the week that we do the seven times table, then 'eight sevens' when we do the eight times table. (p. 21)

Askew, M., & Ebbutt, S. (n.d). *The numeracy file*. London: BEAM education.

Rather than just getting 'three free', students can get 'five free' if they also relate division to fractions. The square numbers have less 'free gifts' because they are made up of numbers multiplied by themselves.

An Approach for Older Students

There are many students in upper primary and lower secondary school who lack fluency with basic multiplication facts. When students reach this stage, they become self conscious of their lack of fluency. Often these students have developed negative attitudes toward mathematics in general and the basic multiplication facts in particular.

The following approach draws on the ideas contained in the section, "Reducing the Memory Load" [p. 12]. In addition, some novel approaches to learning the tables, such as finger methods are suggested.

It is helpful for students to have a copy of a multiplication grid [See p. 21] in front of them as you work through the key ideas associated with reducing the number of facts to be remembered.

Start by examining the entire multiplication grid and acknowledge there is a lot to learn. Explain there are some tricks that will save a lot of time and effort.

- Discuss the multiplication property of zero and then colour in all the 'zero times tables'. Make brief reference to the fact that 3 x 0 and 0 x 3, both give the same result of zero. Later the commutative property can be discussed in greater detail.
- Discuss the multiplication property of one and then colour in all the 'one times table', that is both 1 x ... and ... x 1.
- Most students do not experience trouble counting in twos or skip counting in twos. If you can count in two as far as 20, then you know the 'two times table'. Ask the students to colour in all the 'two times table'.
- Count in fives and discuss the 5, 0, 5, 0 pattern that is formed. Colour in the 'five times table'.

- Discuss multiplying by ten and ask the students to describe the pattern they observe. Discourage the suggestion that when you multiply a number by ten you 'add a zero' as it is not only misleading, breaking down when you multiply a decimal number by ten (0·5 x 10), but it is also incorrect. Colour in all the 'ten times table'. [Note: While the 'ten times table' is not really a basic multiplication fact, the patterns make it worthwhile considering. The 'ten times tables' are simple to learn and give students a sense of achievement that they 'know' so many facts.]
- Focus on the commutative property. **IF students understand the commutative property, THEN they have far less facts to remember.** Note: this is a big 'IF' and it is worth taking the time to establish the commutative property of multiplication as it saves so much time and effort in the long run. Encourage the students to colour in the facts that are represented. The students may notice that these facts fall one side of the square numbers.
- The figure below shows the remaining facts to be learned. These include those in white boxes and those in black boxes – the square numbers.
- This leaves twenty-one facts to be learned, six of which are the square numbers. These are shown above right.

x	0	1	2	3	4	5	6	7	8	9	10
0	0	0	0	0	0	0	0	0	0	0	0
1	0	1	2	3	4	5	6	7	8	9	10
2	0	2	4	6	8	10	12	14	16	18	20
3	0	3	6	9	12	15	18	21	24	27	30
4	0	4	8	12	16	20	24	28	32	36	40
5	0	5	10	15	20	25	30	35	40	45	50
6	0	6	12	18	24	30	36	42	48	54	60
7	0	7	14	21	28	35	42	49	56	63	70
8	0	8	16	24	32	40	48	56	64	72	80
9	0	9	18	27	36	45	54	63	72	81	90
10	0	10	20	30	40	50	60	70	80	90	100

3 x 3 = 9
4 x 3 = 12
6 x 3 = 18
7 x 3 = 21
8 x 3 = 24
9 x 3 = 27

4 x 4 = 16
6 x 4 = 24
7 x 4 = 28
8 x 4 = 32
9 x 4 = 36

6 x 6 = 36
7 x 6 = 42
8 x 6 = 48
9 x 6 = 54

7 x 7 = 49
8 x 7 = 56
9 x 7 = 63

8 x 8 = 64
9 x 8 = 72

9 x 9 = 81

These twenty-one table facts may be grouped in various ways. For example, the square numbers:

3 x 3 = 9
4 x 4 = 16
6 x 6 = 36
7 x 7 = 49
8 x 8 = 64
9 x 9 = 81

Students can be encouraged to draw the arrays associated with the square numbers and note that a square is formed. They will note how the factors are the same also [See 'Square Numbers', p. 48].

The three times table is another grouping that may be explored. The multiples of three may be examined on a number grid. Likewise the digit-sum pattern that is formed when the digits of the products are added may be examined [See 'x 3', p. 40].

3 x 3 = 9
3 x 4 = 12
3 x 6 = 18
3 x 7 = 21
3 x 8 = 24
3 x 9 = 27

The nine times table forms another group that may be considered.

3 x 9 = 27
4 x 9 = 36
6 x 9 = 54
7 x 9 = 63
8 x 9 = 72
9 x 9 = 81

Various methods including using patterns and finger methods may be used to support the learning of the 'nine times table'. These ideas are outlined in the activities section.

Finding Out Which Facts Students Know

The activities and ideas contained on the following pages are designed to help you find out what basic multiplication facts the students know.

Making a Tables Map (p. 21)

Rather than assume a student can or can't recall certain multiplication facts, an effort should be made to determine what is known so that instruction can focus on areas of weakness. The idea behind the table map activity is to map students' table knowledge. This map may be used to diagnose problems students have with tables and help choose activities suited to the needs of individual students. To make a Table Map to find out what table facts children know, you will need:

- a tables chart
- two coloured pencils or pencils
- two different coloured, ten sided dice, (let the zero represent ten).

Roll the two dice to generate tables questions. Always refer to one colour first and the other colour second. This will allow you to determine whether the child has developed an understanding of the commutative property (i.e. 3 x 8 is the same as 8 x 3).

Roll the two dice and call out the numbers that show. If the response is instantaneous, colour the appropriate fact on the chart using one colour. If the child takes a little time to determine the answer, then use a different colour to shade the fact. If the child cannot answer the question, leave the fact blank.

Once most facts have been 'tested' using the dice, any remaining facts may be checked orally. **The 'map' should now show which facts are known, which can be worked out and those that require work.** Activities that focus on the unknown facts may be used to improve knowledge of the tables facts.

The Paper Calculator (p. 22 - 23)

The Paper Calculator is based on the work of John Napier (1550 – 1617) who devised a system, later known as Napier's Rods or Bones, which enabled large multiplications to be performed with ease. The set of strips from 0 to 9 represent all of the table facts.

If students are encouraged to make their own rods they will gain valuable practice in the table facts and have the opportunity to notice patterns within. While the students are completing each of the rods teachers may observe how the students go about completing them; whether they hesitate or make mistakes. The completed rods may be checked for mistakes. A record may be kept of the tables that are causing difficulties. To successfully perform a calculation using the rods required an understanding of place value and recognising the evolving pattern.

Triangular Tables (p. 24 - 25)

The triangular table chart is designed so that no repeated multiplication facts are included. While this activity will not indicate whether students can make use of the commutative property of multiplication it will show you which multiplication facts the students know and which they don't know. As such it would make an ideal work sample.

I Know it, I can work it out, I have no idea (p. 26)

This activity is designed to be used as a self check, where students decide on which facts they know fluently and which ones they use strategies to work out, that is, take longer to calculate, and those they do not know.

The 68 questions (4 columns of 17) shown in the table (on page 26) represent all the basic facts from the 'two times table' to the 'nine times table'. Two 'zero times facts' and a 'one times fact' have been included so teachers can check whether individual students have grasped the multiplication properties of zero and one. Two 'ten times' facts have been included to check whether students have discerned to the pattern associated with multiplying by ten. All the square numbers from 2 x 2 to 9 x 9 have been included along with the other facts and their commutative partners. For example, 6 x 4 and 4 x 6. While time should not be an issue (see earlier comment on timed tests) a student who *knows* all the facts will be able to complete a column very quickly.

Self Check (p. 27)

This is another self-diagnostic tool that considers other factors that might be having a bearing on a student learning the basic multiplication facts. For example, if a student does not know that the word product means multiply then when asked to find the product of 7 and 8 he or she may not give an answer despite knowing the answer to 7 x 8. The student's real ability is thus being masked.

Making A Tables Map

Name: ______________________________

Date: __________

x	0	1	2	3	4	5	6	7	8	9	10
0	0	0	0	0	0	0	0	0	0	0	0
1	0	1	2	3	4	5	6	7	8	9	10
2	0	2	4	6	8	10	12	14	16	18	20
3	0	3	6	9	12	15	18	21	24	27	30
4	0	4	8	12	16	20	24	28	32	36	40
5	0	5	10	15	20	25	30	35	40	45	50
6	0	6	12	18	24	30	36	42	48	54	60
7	0	7	14	21	28	35	42	49	56	63	70
8	0	8	16	24	32	40	48	56	64	72	80
9	0	9	18	27	36	45	54	63	72	81	90
10	0	10	20	30	40	50	60	70	80	90	100

Date: __________

x	0	1	2	3	4	5	6	7	8	9	10
0											
1											
2											
3											
4											
5											
6											
7											
8											
9											
10											

Making a Paper Calculator

To make your own "paper calculator" you need to construct a set of 10 strips containing all the tables.

The '5 x strip' has been done for you. Notice how the tens part of the answer and the units, or ones, are separated. The tens digit goes above the diagonal line and the ones digit below the line. Complete the rest.

Cut them out vertically, down each strip.

x	0	1	2	3	4	5	6	7	8	9
1	0/0	0/1	0/2	0/3		0/5				
2	0/0	0/2	0/4	0/6		1/0				
3	0/0	0/3	0/6			1/5				
4	0/0	0/4				2/0				
5	0/0	0/5				2/5				
6	0/0	0/6				3/0				
7	0/0	0/7				3/5				
8	0/0	0/8				4/0				
9	0/0	0/9				4/5				

The paper calculator is based on Napier's Rods. John Napier (1550 – 1617) developed his calculating system long before electronic calculators were invented.

Using a Paper Calculator

The rods can be used to calculate in the following manner.

e.g. 6 x 549

Pick out the 4, 5 and 9 rods and place them in correct order (5, 4 and then 9).

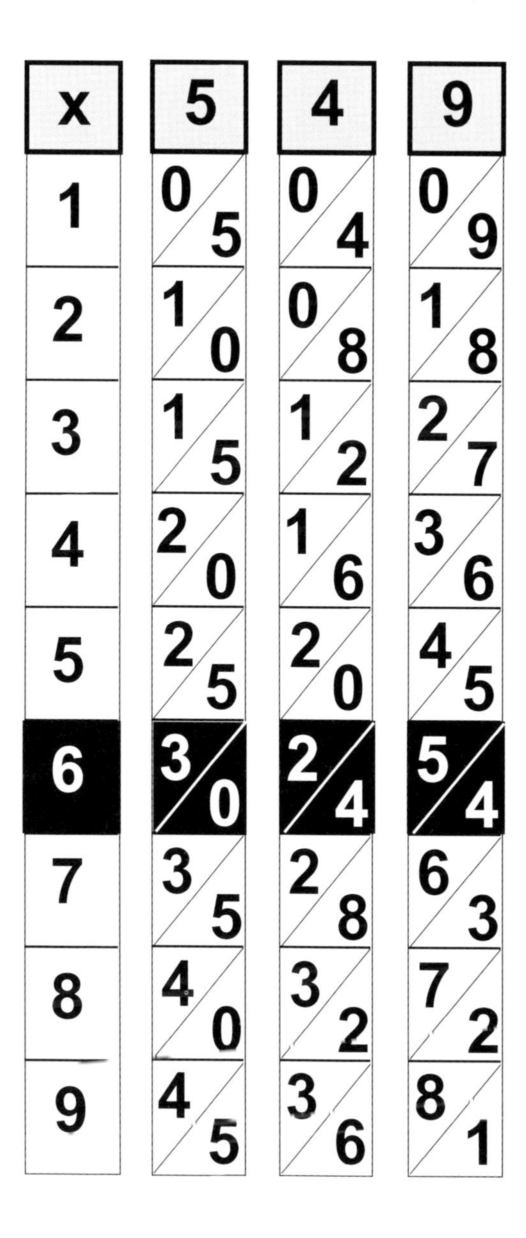

The answer to 6 x 549 is given in row 6. Simply add the numbers between the diagonal lines.

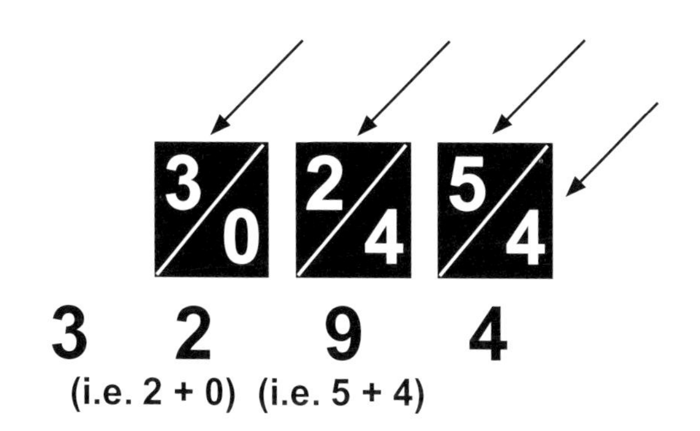

Use the rods to calculate 8 x 407.

Note: if a digit is repeated in a number such as 644, then you will need to borrow an extra four rod to complete the calculation.

Triangular Tables

The Triangular Tables Chart has been included to highlight the commutative property of multiplication, that is, the result of multiplying 3 by 8 is the same as when 8 is multiplied by 3. Traditional tables charts look like this, or like this:

x	0	1	2	3
0	0	0	0	0
1	0	1	2	3
2	0	2	4	6
3	0	3	6	9

1 x 2 = 2	1 x 3 = 3
2 x 2 = 4	2 x 3 = 6
3 x 2 = 6	3 x 3 = 9
4 x 2 = 8	4 x 3 = 12
5 x 2 = 10	5 x 3 = 15
6 x 2 = 12	6 x 3 = 18
7 x 2 = 14	7 x 3 = 21
8 x 2 = 16	8 x 3 = 24
9 x 2 = 18	9 x 3 = 27

Each of these methods involves writing the fact twice. For example 2 x 3 and 3 x 2 are repeated. With a triangular tables chart you only have to write each table fact once! A copy of the triangular tables chart is provided for students to complete. As the students complete the Triangular Tables Chart, focus on the commutative property of multiplication.

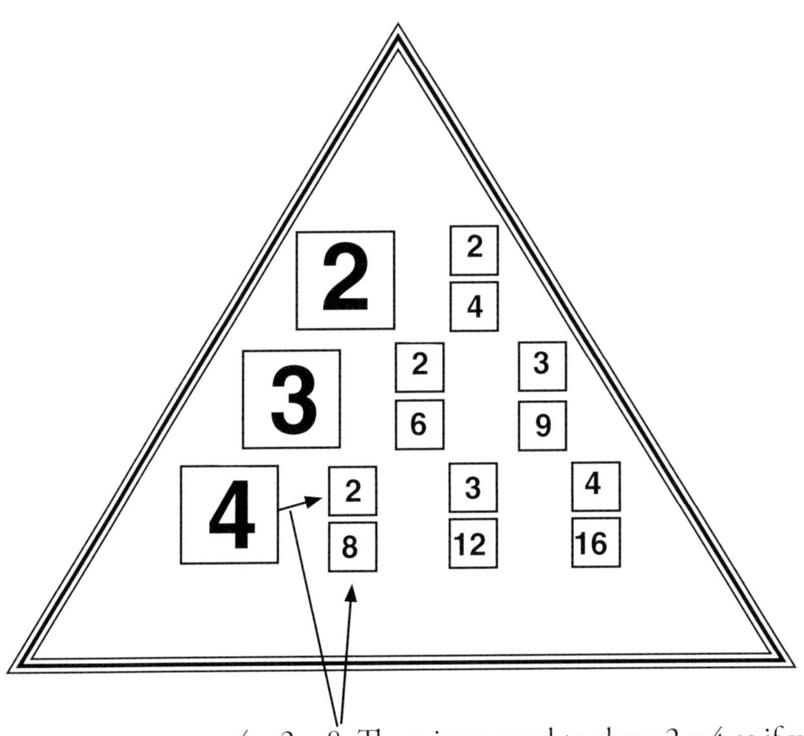

4 x 2 = 8. There is no need to show 2 x 4 as if you know 4 x 2 then you should know the result of multiplying 2 x 4. This is reflected at the top of the triangle.

Triangular Tables

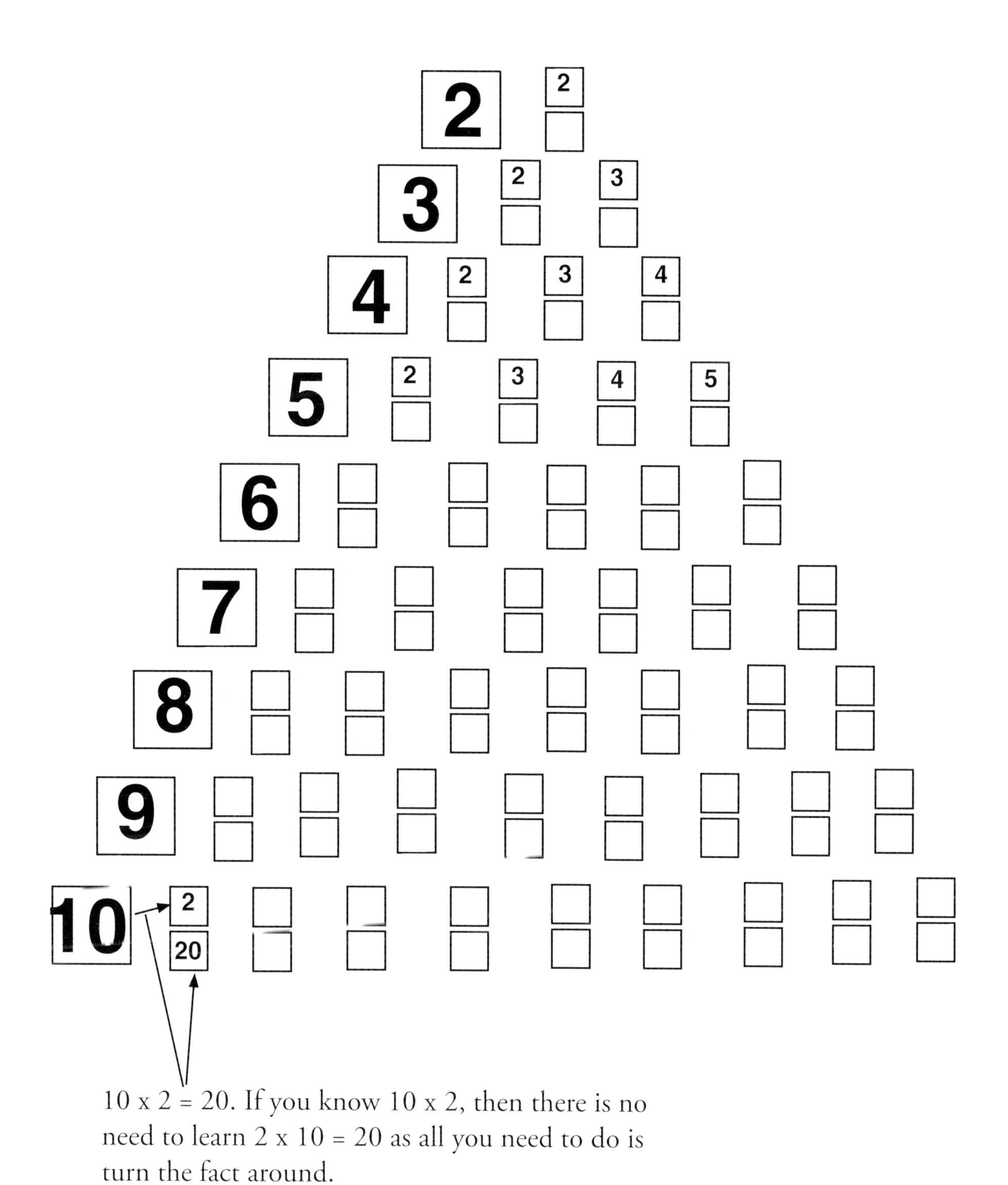

10 x 2 = 20. If you know 10 x 2, then there is no need to learn 2 x 10 = 20 as all you need to do is turn the fact around.

I Know it, I Can Work it Out, I Have no Idea

Complete the table below. Write the answers to the calculations you know in *blue*. Write the answers for calculations you need to think about in *red*. Leave *blank* any that you don't know.

Name: **Date:**

3 x 3 =	7 x 2 =	4 x 3 =	3 x 5 =
6 x 2 =	3 x 4 =	5 x 2 =	4 x 4 =
7 x 3 =	8 x 4 =	2 x 9 =	6 x 7 =
9 x 4 =	6 x 6 =	3 x 8 =	7 x 4 =
0 x 8 =	3 x 10 =	5 x 5 =	8 x 6 =
5 x 6 =	9 x 9 =	7 x 8 =	8 x 3 =
8 x 8 =	4 x 6 =	9 x 6 =	4 x 9 =
6 x 9 =	8 x 5 =	2 x 5 =	7 x 9 =
1 x 7 =	4 x 0 =	2 x 4 =	8 x 2=
7 x 7 =	2 x 2 =	7 x 6 =	6 x 5 =
5 x 7 =	4 x 5 =	3 x 2 =	3 x 6 =
6 x 3 =	9 x 7 =	3 x 7 =	5 x 8 =
9 x 5 =	6 x 8 =	6 x 4 =	5 x 3 =
2 x 6 =	4 x 8 =	5 x 9 =	8 x 9 =
8 x 7 =	4 x 7 =	2 x 8 =	4 x 2 =
2 x 3 =	9 x 2 =	7 x 5 =	9 x 8 =
9 x 3 =	5 x 4 =	3 x 9 =	2 x 7 =

Self Check

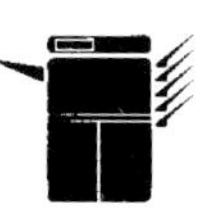

Name: ______________________ **Date:** __________

The following self check is designed to help you spot your strengths and weaknesses when it comes to learning the tables.

I can count in:

- ❑ twos ______________________
- ❑ threes ______________________
- ❑ fours ______________________
- ❑ fives ______________________
- ❑ sixes ______________________
- ❑ sevens ______________________
- ❑ eights ______________________
- ❑ nines ______________________
- ❑ What is another way of saying 7 + 7 + 7 + 7? __________
- ❑ List three **multiples** of nine. __________
- ❑ List the **factors** of 36. __________
- ❑ What is the **product** of 8 and 3? __________

You will need a piece of paper to write down the answers to the following questions.

- ❑ Draw a picture to show 3 x 4.
- ❑ Describe any patterns that you use to help you remember certain tables.
- ❑ Explain how multiplication is related to division. Use an example to show how multiplication is related to division.
- ❑ If you already know that 8 x 3 is 24, what else do you know or can be able to work out?
- ❑ List three table facts that you find easy to remember. Explain why you think they are easy to remember.
- ❑ List three table facts that you find hard to remember. Explain why you think they are hard to remember.

Why do Some Students Have Difficulty Learning the Tables?

Students appear to have little difficulty learning sporting and entertainment facts and trivia. Students also learn to read and spell hundreds of words and yet they struggle to remember the basic multiplication facts. Clearly for the most part while memory does play a role in learning the basic multiplication facts, it is not the most important factor and should not be blamed for students not learning them. It is acknowledged, however, that there are some students with poor memories that will need to rely on using strategies to reconstruct the basic multiplication facts.

Experience suggests that students who rely solely on memorising the basic multiplication facts are making the job of learning the 'tables' much harder than it needs to be. Consider the symbols shown in the grid below. Imagine trying to remember each individual symbol. It would certainly be difficult. However, on deeper examination of the grid certain patterns become obvious that making the job of learning the symbols much simpler. For example, consider the first row and the first column – the symbols are all the same. Likewise focus on the second row and column. Now look at the sixth column and the sixth row.

Clearly looking for, and using *patterns* will assist students to reduce the number of table facts that need to be memorised. Further examination of the symbols in the grid will indicate that the symbols are repeated on either side of the diagonal that runs from the top left of the grid to the bottom right. Consider the last column and row. What pattern can you spot?

Many of the activities contained in the Activities section focus on the patterns to be found in the multiplication grid. Students are less likely to notice the patterns when the basic multiplication facts are printed in the traditional 'table format' [See right of p. 6].

	♣	♦	♥	♠	❖	✜	✫	★	☆	✪	❍
♣	♣	♣	♣	♣	♣	♣	♣	♣	♣	♣	♣
♦	♣	♦	♥	♠	❖	✜	✫	★	☆	✪	❍
♥	♣	♥	❖	✫	☆	♦♣	♦♥	♦❖	♦✫	♦☆	♥❍
♠	♣	♠	✫	✪	♦♥	♦✜	♦☆	♥♦	♥❖	♥★	♠❍
❖	♣	❖	☆	♦♥	♦✫	♥♣	♥❖	♥☆	♠♥	♠✫	❖❍
✜	♣	✜	♦♣	♦✜	♥♣	♥✜	♠♣	♠✜	❖♣	❖✜	✜❍
✫	♣	✫	♦♥	♦☆	♥❖	♠♣	♠✫	❖♥	❖☆	✜❖	✫❍
★	♣	★	♦❖	♥♦	♥☆	♠✜	❖♥	❖✪	✜✫	✫♠	★❍
☆	♣	☆	♦✫	♥❖	♠♥	❖♣	❖☆	✜✫	✫❖	★♥	☆❍
✪	♣	✪	♦☆	♥★	♠✫	❖✜	✜❖	✫♠	★♥	☆♦	✪❍
❍	♣	❍	♥❍	♠❍	❖❍	✜❍	✫❍	★❍	☆❍	✪❍	♦❍❍

The Development Phase

Earlier reference was made to the following model. It depicts a transition phase between using strategies to derive new facts that eventually become part of the bank of 'known facts'.

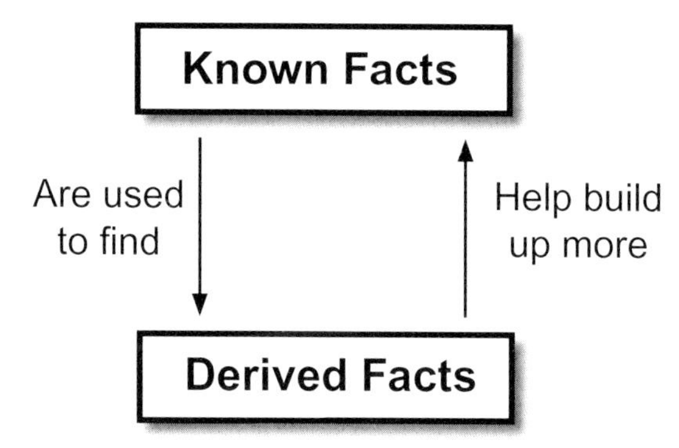

Once facts become 'known' drill and practice activities may be used to increase the speed of recall. Drill and practice activities designed to improve fluency with the basic multiplication facts is the subject of the last section of this book. Prior to this, however, considerable consolidation needs to take place.

Building a Set of Known Facts

This section contains ideas and activities for developing patterns and connections between various basic multiplication facts. Many of the activities feature the properties of multiplication and strategies outlined in the first section of this book.

Encourage students to discuss how the various strategies and properties help to consolidate their bank of facts.

The Role of Language

During discussions about multiplication, and the various strategies for multiplying emphasise the language component. ***Many students who experience difficulty with the basic multiplication facts are actually experiencing difficulty with the language associated with multiplication.*** For example, a student might give the correct response to multiplying 4 by 5, but not have a clue when asked to find the *product* of 4 and 5.

The expression 'three times four' is not all that helpful as it does not really describe the multiplication concept. The expression "lots of" tends to focus on the repeated addition concept of multiplication. It is probably simpler to use an expression like '3 *fours* are 12' or '3 multiplied by 4', as it may be applied to repeated groups or arrays.

When discussing strategies such as doubling, expressions such as "twice" may also come up. Acknowledge that twice 5 gives the same result as double 5.

Make the Links

Whenever possible, draw links to related table facts. If students have learned a family of facts that relate to one another they are less likely to forget them.

Emphasise Strategies

Many students will make use of mental strategies without really thinking about what they are doing. Ask them to explain the strategy to use and help them understand why it works. Verbalising will assist students to better understand the strategy they are using.

Consider the property

As discussed earlier, many properties of the number system underpin work with the basic multiplication facts. Show how these make the job of learning the basic multiplication facts easier. In particular, how the commutative property reduces the number of facts that need to be learned.

Number Lines

The number line is one model that is used to help children learn to multiply. ***It is closely related to the repeated addition model of multiplication.*** Consider the following example. The students would be told to jump along the number line in threes.

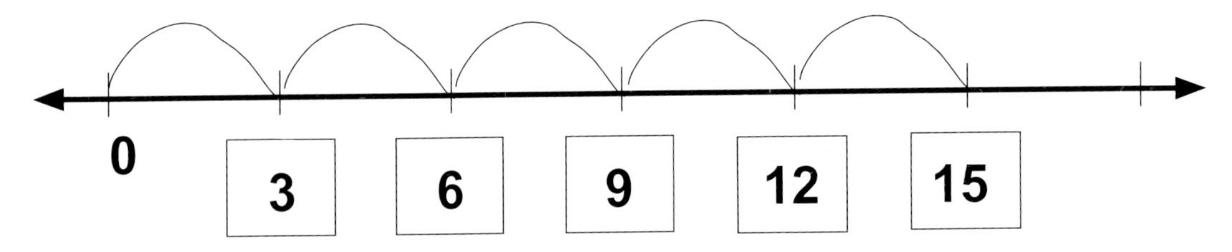

After each jump the students could fill in the box to show the next multiple of three. Once five jumps have been completed, the count (15) should give the result for 5 x 3.

It is important that students learn to write the appropriate number sentence to match the situation. In this case 5 x 3 is 3 + 3 + 3 + 3 + 3, which is 15. The students can write:

$$5 \times 3 = 3 + 3 + 3 + 3 + 3$$

This will help emphasise that the 'equals symbol' represents equality and the two sides of the equation are balanced. Writing:

3 + 3 + 3 + 3 + 3 = 15 and

5 x 3 = 15 are equally true.

Links may be made to the commutative property of multiplication which explains that the order in which a multiplication is performed does not alter the result. Using a number line helps to illustrate that, while the result may be the same, the situation differs. For example, 3 x 5 would look like this.

The number sentences associated with this number line (see below) would be :

3 x 5 = 5 + 5 + 5 or

5 + 5 + 5 = 15 or

3 x 5 = 15.

Notice how the repeated addition concept of multiplication is emphasised using a number line. Note, too, how the commutative property is demonstrated and how different the situations appear on the number line model.

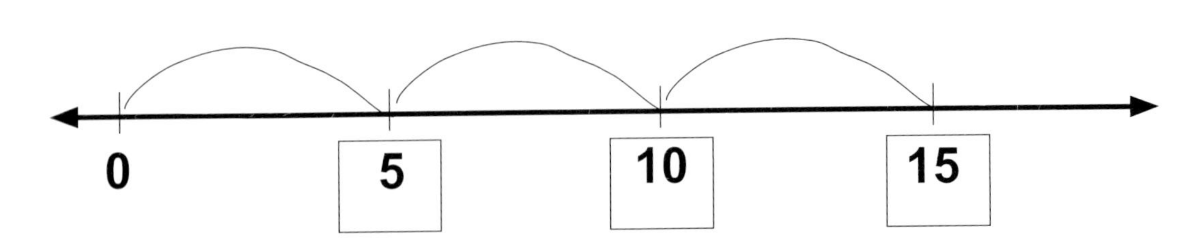

Making a Pocket Multiplier

The pocket multiplier is a simple aid that students can make. ***It emphasises the repeated addition nature of multiplication.*** The instructions for making a pocket multiplier are given below.

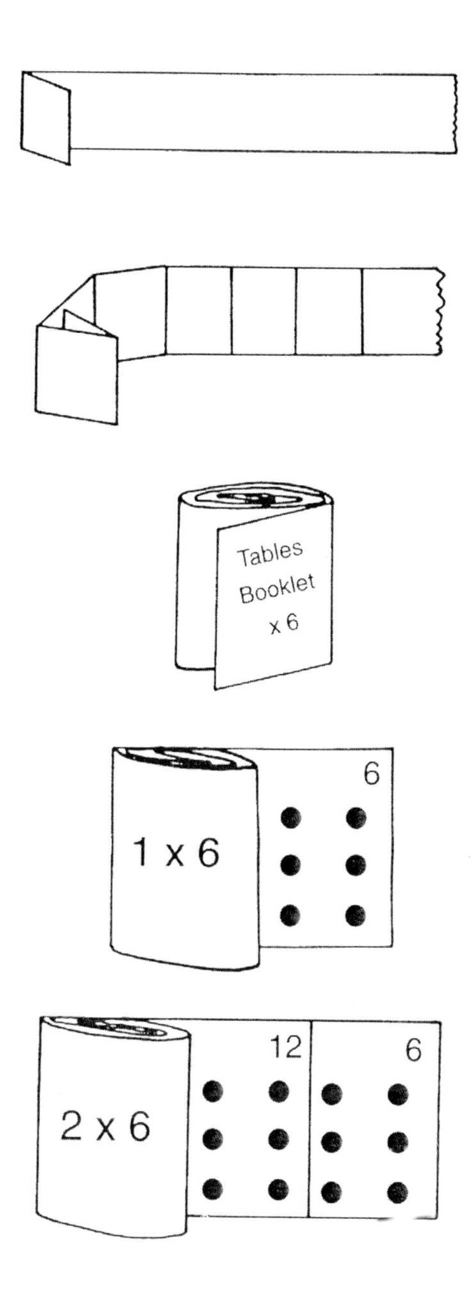

You will need to cut a strip of paper 55 cm long and 7 cm wide.

The strip may then be folded wrap-around style every 4 - 5 cm. [Note: As the pocket multiplier is folded the size of the rectangles needs to be increased. Keep the 1, 2, 3 & 4 rectangles the same size. Add 3 mm to the 5, 6 & 7 rectangles and a further 2 mm to the 8, 9 & 10 rectangles. This should assist with the folding.] Continue to fold the strip for its entire length. Cut off any surplus paper so as to form a neat rectangle.

Choose a multiplier. This could be based on a table fact that the student needs to learn. In this example the pocket multiplier shows the multiples of six. Write this on the front of the pocket multiplier.

Turn back the front page and on the left side write '1 x 6' and on the right make a pattern of 6 dots. Write the number 6 in the top right hand corner.

Open the strip of paper one more page, write 2 x 6 on the left side, make another pattern of 6 dots and write the number 12 in the top right corner.

Continue to open up the strip of paper and each time add one more multiple of 6. Notice how repeated addition is modelled (6 + 6 + 6 + 6 + 6). Students may use the pocket multiplier to test themselves and periodically review specific table facts.

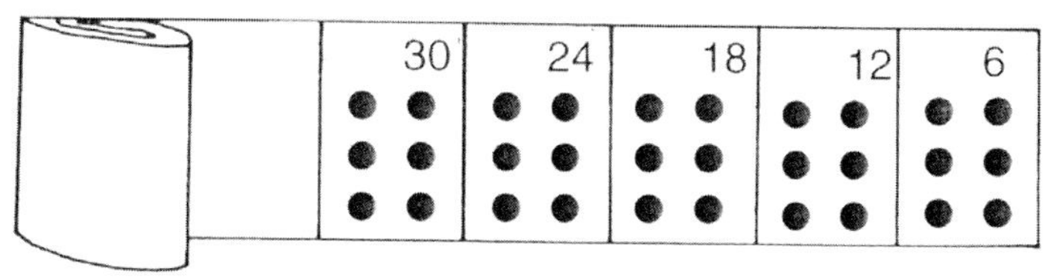

The Array Game

As the name suggests, this activity helps to develop the array concept of multiplication. ***The array model is ideal for developing the commutative property of multiplication.***

You will need two six-sided dice and 5 mm grid paper.

Two dice are thrown to produce the dimensions of a rectangle. For example, if the numbers shown are 3 and 4 then a rectangle consisting of three rows of four is drawn. Students draw in the appropriate boundary and then write in the appropriate number sentence. In this case the student would write 3 x 4, representing that the diagram shows three rows of four.

3 x 4

Students who do not know the result of multiplying 3 by 4 can count squares to 'work out' the answer. Once students are familiar with the idea of drawing arrays then a game may be played between two students. Each player uses a different colour to capture territory on a grid (see right).

Rules

- One player rolls the dice and colours in a rectangle, matching the dimensions shown on the dice, on the grid. If the numbers 3 and 4 are shown by the dice then the player may shade in a 3 x 4 or a 4 x 3 rectangle on the grid.
- The second player then rolls the dice and shades in an appropriate rectangle on the grid.
- If a player cannot draw a rectangle in the space available then he/she misses a turn.
- The player who has covered the most squares after ten rounds is the winner.

Variation

Use two, ten-sided dice.

After writing the appropriate number sentence in the array, students can be encouraged to write down the associated facts that are in the same family. For example, in the case of the 3 x 4 array, the students would write down 4 x 3, 12 ÷ 3 = 4 and 12 ÷ 3 = 4. Links to fractions may be encouraged by writing the associated fraction facts, $^1/_4$ of 12 is 3 and $^1/_3$ of 12 is 4.

Extending the Game

Arrays are ideal for illustrating the power of the ***doubling strategy***.

The game may be altered to incorporate the use of the doubling strategy. This time after throwing the dice and drawing the related rectangle the students have to double one of the numbers and draw the new rectangle that is formed.

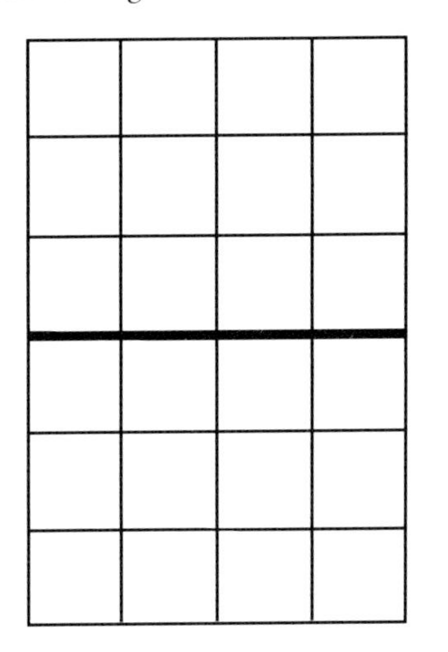

Ask the students to 'work out' the new answer.

Draw the students' attention to the way in which the doubling strategy may be used to derive a new fact from a previously known fact. That is, double 3 x 4 produces the result of multiplying 6 by 4. The students could also explore what happens when the other dimension (factor) is doubled, that is 8 – to produce the result to the new fact 3 by 8.

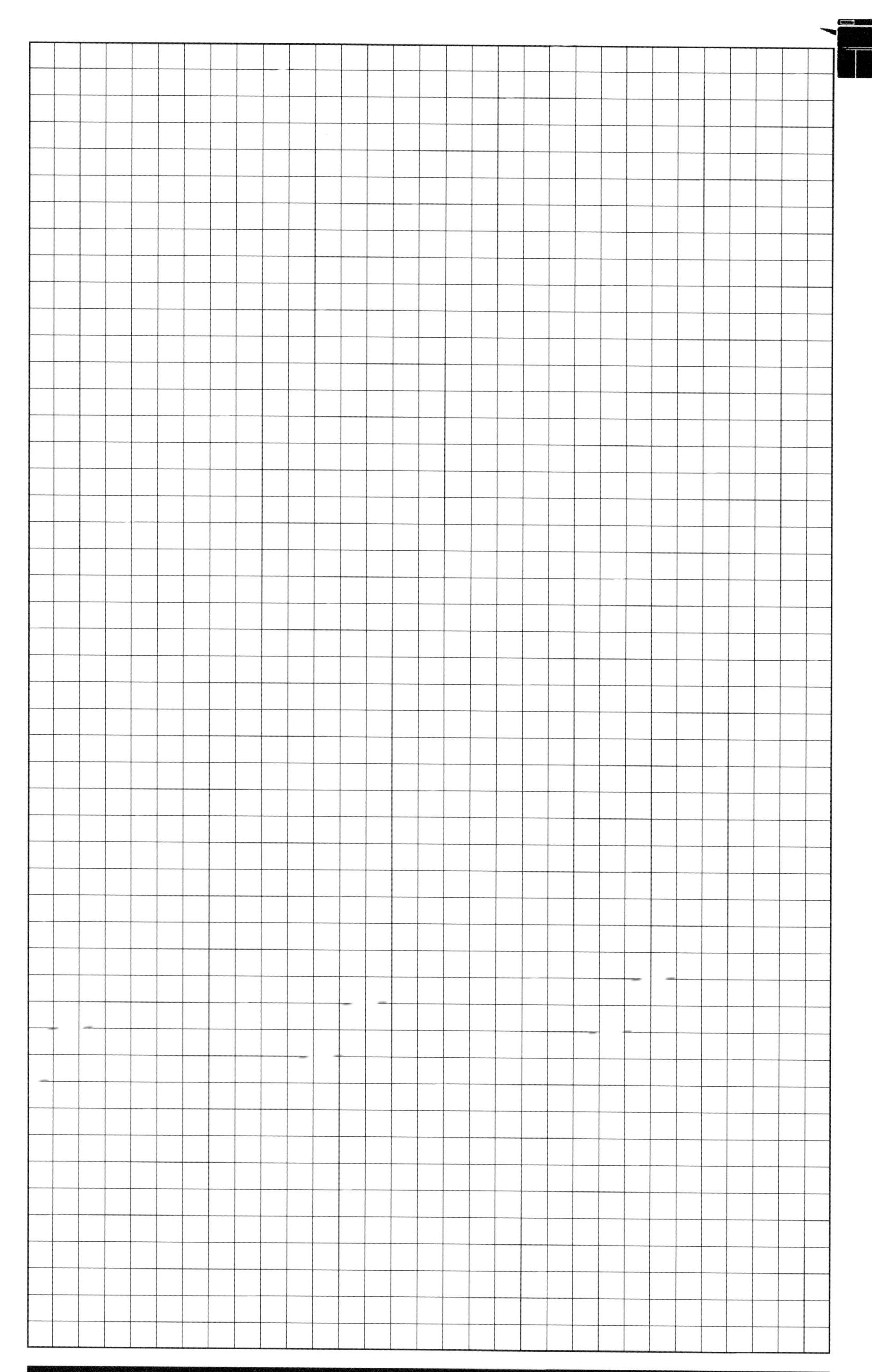

Array Cover Ups

Background

Earlier reference was made to the array model of multiplication. The previous activity "The Array Game" made use of that model and built upon it by extending it to incorporate the doubling strategy.

In this activity the array model, and the visual image associated with it, are used to consolidate the basic multiplication facts. Through the use of the array model, links can be made to factors and products. Links may be made to division also.

In the diagram shown on the right, students may count individual dots or rows of dots to reach a total of twenty-one. Students could be asked to count the number of dots in a row and the number of dots in a column and look for links. The 3 and the 7 are factors of 21 (which is called the product).

Further discoveries may be made by changing the paper used to cover the array. Transparent paper will provide you with the opportunity to explore different strategies and to consider complements of 100.

Changing the array to include numbers on the outside will formalise the relationship between the factors and the product [See p. 36]. Try starting with the square numbers first as the relationship is more obvious.

Provide a 10 x 10 dot array for each student [See p. 35]. At this point, avoid using the numbered array shown on page 36 as the links between the numbers in the columns and rows will be drawn from the students' observations.

Discuss the make-up of the array. Some students will count dots, others will note the array is ten dots wide and ten dots high and work out the total number of dots as being 100.

Using an overhead transparency of the array, cover various parts of the array and ask related questions. For example, cover all but the top row or the top two rows and ask, "How many dots?" Cover all but the first three columns and ask "How many dots?" Continue asking similar questions until the students become fluent in determining the number of dots.

Next cover some rows and columns (similar to the diagram shown below) and ask, "How many dots are showing?" Ask students to explain how they arrived at the answer. Some students may have counted the number of dots in each row and then skip counted. Using the example below, this would be 3, 6, 9, 12, 15, 18, 21. Other students may think of seven rows of three and arrive at the result of 21. If the paper used to cover the array is transparent, then some students may begin with 3 x 10 and then subtract the 9 dots that are covered, but can be seen through the paper covering.

100 Dot Arrays

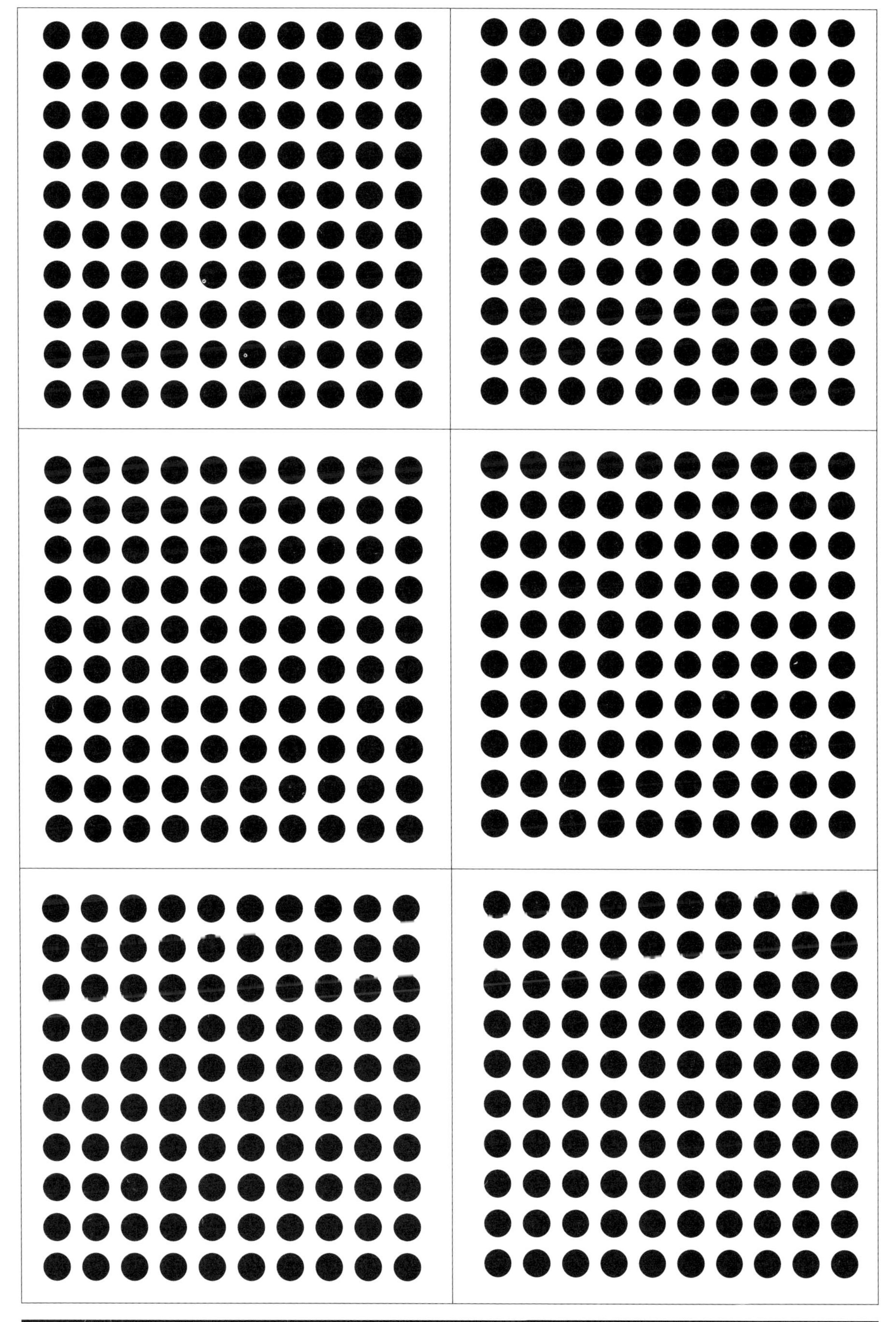

Multiplication Dot Arrays

x	1	2	3	4	5	6	7	8	9	10
1	●	●	●	●	●	●	●	●	●	●
2	●	●	●	●	●	●	●	●	●	●
3	●	●	●	●	●	●	●	●	●	●
4	●	●	●	●	●	●	●	●	●	●
5	●	●	●	●	●	●	●	●	●	●
6	●	●	●	●	●	●	●	●	●	●
7	●	●	●	●	●	●	●	●	●	●
8	●	●	●	●	●	●	●	●	●	●
9	●	●	●	●	●	●	●	●	●	●
10	●	●	●	●	●	●	●	●	●	●

x	1	2	3	4	5	6	7	8	9	10
1	●	●	●	●	●	●	●	●	●	●
2	●	●	●	●	●	●	●	●	●	●
3	●	●	●	●	●	●	●	●	●	●
4	●	●	●	●	●	●	●	●	●	●
5	●	●	●	●	●	●	●	●	●	●
6	●	●	●	●	●	●	●	●	●	●
7	●	●	●	●	●	●	●	●	●	●
8	●	●	●	●	●	●	●	●	●	●
9	●	●	●	●	●	●	●	●	●	●
10	●	●	●	●	●	●	●	●	●	●

x	1	2	3	4	5	6	7	8	9	10
1	●	●	●	●	●	●	●	●	●	●
2	●	●	●	●	●	●	●	●	●	●
3	●	●	●	●	●	●	●	●	●	●
4	●	●	●	●	●	●	●	●	●	●
5	●	●	●	●	●	●	●	●	●	●
6	●	●	●	●	●	●	●	●	●	●
7	●	●	●	●	●	●	●	●	●	●
8	●	●	●	●	●	●	●	●	●	●
9	●	●	●	●	●	●	●	●	●	●
10	●	●	●	●	●	●	●	●	●	●

x	1	2	3	4	5	6	7	8	9	10
1	●	●	●	●	●	●	●	●	●	●
2	●	●	●	●	●	●	●	●	●	●
3	●	●	●	●	●	●	●	●	●	●
4	●	●	●	●	●	●	●	●	●	●
5	●	●	●	●	●	●	●	●	●	●
6	●	●	●	●	●	●	●	●	●	●
7	●	●	●	●	●	●	●	●	●	●
8	●	●	●	●	●	●	●	●	●	●
9	●	●	●	●	●	●	●	●	●	●
10	●	●	●	●	●	●	●	●	●	●

x	1	2	3	4	5	6	7	8	9	10
1	●	●	●	●	●	●	●	●	●	●
2	●	●	●	●	●	●	●	●	●	●
3	●	●	●	●	●	●	●	●	●	●
4	●	●	●	●	●	●	●	●	●	●
5	●	●	●	●	●	●	●	●	●	●
6	●	●	●	●	●	●	●	●	●	●
7	●	●	●	●	●	●	●	●	●	●
8	●	●	●	●	●	●	●	●	●	●
9	●	●	●	●	●	●	●	●	●	●
10	●	●	●	●	●	●	●	●	●	●

x	1	2	3	4	5	6	7	8	9	10
1	●	●	●	●	●	●	●	●	●	●
2	●	●	●	●	●	●	●	●	●	●
3	●	●	●	●	●	●	●	●	●	●
4	●	●	●	●	●	●	●	●	●	●
5	●	●	●	●	●	●	●	●	●	●
6	●	●	●	●	●	●	●	●	●	●
7	●	●	●	●	●	●	●	●	●	●
8	●	●	●	●	●	●	●	●	●	●
9	●	●	●	●	●	●	●	●	●	●
10	●	●	●	●	●	●	●	●	●	●

Doubling and Halving

Doubling and halving is an extremely powerful strategy that is not only helpful when it comes to learning the basic multiplication facts but also for multiplying and dividing with larger numbers.

Once a few basic multiplication facts are known these may be used in conjunction with the doubling strategy to produce further multiplication facts. For example, if you know your 'two times table', the 'four times table' may be generated using doubling.

3 x 2 = 6, doubled is 12, so 3 x 4 is 12 and

7 x 2 = 14, doubled is 28, so 7 x 4 is 28.

Structured teaching will be required in order for this pattern to be established. There are many variations of the doubling strategy that are built upon this understanding.

To multiply by 6, you can multiply by 3 and then double the answer. A similar approach may be used to 'work out' or derive other unknown basic multiplication facts.

Double Double

Doubling may also be used to generate multiplication facts from scratch. For example if asked to calculate 4 x 7 a students could start with 7, then double, to produce 14 and then double again to arrive at the answer of 28. In this case the commutative property of multiplication comes into play because the 7 becomes the focus. The 4 is the 'trigger' that may encourage a student to think of using the doubling strategy. Similar reasoning may be used with larger numbers.

Double Double Double

The 'double double double' strategy is a good way to multiply by eight. For example 7 x 8 may be thought of as:

double 7 is 14

double 14 is 28

double 28 is 56

so 7 x 8 is 56.

Alternatively a student who knows 4 x 7 and hence 7 x 4 may apply the doubling strategy to 'work out' the result. That is:

7 x 4 = 28, doubled gives 56, which is the answer to 7 x 8.

Halving

Closely allied with doubling is halving. Halving is a useful strategy for dividing by 4 and 8. For example, to calculate 48 ÷ 4 a student may think half of 48 is 24 and half of 24 is 12. To work out 48 ÷ 8 would simply require halving one more time.

Doubling and halving combined

The two approaches may be combined to multiply by 5. For example 26 x 5 may be thought of as 26 x 10, which is easy to calculate. The result, 260, is then halved to reach the answer of 130. When multiplying by 5 it is often easier to multiply by ten and then halve the result.

Function machines

Many teachers use function machines to highlight the doubling strategy. A number is fed into a function machine at one end and the 'machine' doubles the number and it comes out the other end. Linking two function machines highlights the double-double strategy. Linking three function machines highlights the double-double-double strategy.

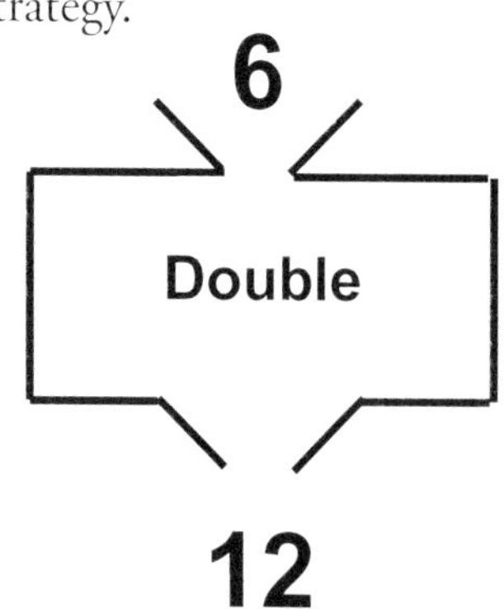

Making the Links

Many students who 'rote learn' their basic multiplication facts do not appreciate the links that exist between the 'tables'. Reference has already been made to students who have not developed an understanding of the commutative property of multiplication. They have not made the link that 8 x 3 = 3 x 8. Failure to make this link means these students have to remember almost twice as many different multiplication facts. All 'table facts', except the square numbers are linked to another fact via the application of the commutative property.

There are, however, many more links or connections that can be made when learning the basic multiplication facts.

Multiplication and division

When learning the basic multiplication facts it makes sense to link multiplication and division. Students who have a good understanding of the operation of multiplication will know that division is the inverse of multiplication and that division will undo multiplication. This link needs to be made explicit as it will make the task of learning the division facts so much simpler.

When learning one fact such as 8 x 3, not only does it make sense to link it to 3 x 8, but also to link it to the associated division facts 24 ÷ 3 = 8 and 24 ÷ 8 = 3. These related facts form a family. The family may be extended to include $^1/_3$ of 24 = 8 and $^1/_8$ of 24 = 3.

Furthering the links

Given all the effort that goes into learning the basic number facts, of which the multiplication and division facts are a part, it makes sense to maximise all the calculations that may be produced from them. For example if you know 8 x 3 = 24, then you should know the answer to a range of related questions such as 8 x 30, 30 x 8, 80 x 30, 0·8 x 30 and so on.

The following ideas are designed to help students foster as many links between various number facts as possible.

Sorting the calculation

Students write the three numbers involved in a basic multiplication fact they know on three separate sheets of paper. For example, the students might write 3, 8 and 24 on to three pieces of paper and hand them to a partner to form four different calculations or number sentences.

Back to the start

A simple format similar to the one shown may be used to highlight the relationship between multiplication and division. A number is placed in the top box and then the same number is written in the box containing the multiplication and division signs. Students will find that they return to the number they started with.

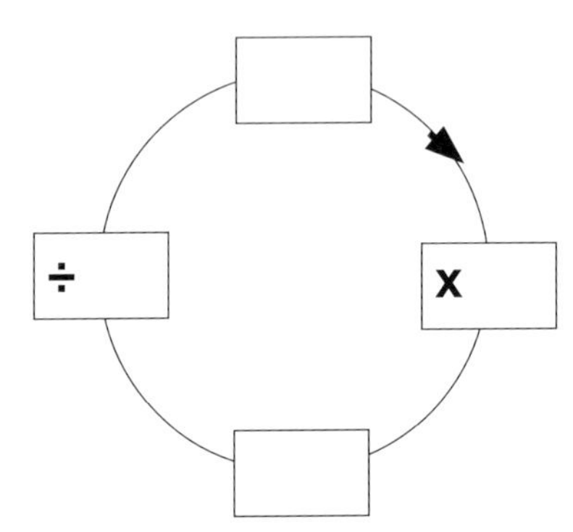

Function machines may be linked to produce a similar result. Passing a number through a 'double' function machine then to a 'halve' function machine will lead to the same number exiting the second function machine.

If I know then I also know ...

This is a simple routine that may be applied to any number fact. A fact, along with the answer is written on the board and the students suggest related facts. For example if you wrote:

6 x 5 = 30

on the board, your students might suggest the following related facts: 5 x 6 = 30, 50 x 6 = 300, 6 x 10 = 60, 6 x 0·5 = 3, 30 ÷ 5 = 6, 5 + 5 + 5 + 5 + 5 + 5 = 30 and so on. Ask the students to explain how their new fact is *related* to the original fact.

When completing the activities that follow, be conscious of the links that can be made and make them explicit to the students. Some of the activities such as "Flash Cards" [p. 51] are designed to help students make the links between related 'triples' such as 3, 8 and 24.

x 2

Most students do not experience much trouble with the 'two times table'. This may be due to all the counting in twos (or skip counting) that they do before embarking on learning the multiplication tables.

Where possible, link the multiples of two and then the 'two times table' to real life objects to give a purpose to the count. Many objects come in pairs, for example, shoes and socks. Body parts such as legs, arms, ears and eyes come in twos. Objects common to many students such a bicycles also come in pairs. Objects such as soap and toilet rolls are often sold in packs of two, or 'twin packs'. These objects can form the basis for questions that will require either counting in twos or the more efficient ability to multiply by two. For example, the following question, "There are 9 bicycles in the bike rack, how many wheels altogether?" leads to the need to multiply 2 by 9 or to repeatedly add two, nine times.

Various models such as number lines, number tracks and number grids may be used to develop a more formal link to the 'two times table'. Calculators may be used to assist students to count in twos beyond twenty or thirty. A record may be kept on a number grid such as the one shown. Note that simple calculators have an inbuilt constant function that may be set to make them count. While the keystroke sequences vary from make and model, the following keystrokes will cause most calculators to 'count in twos'

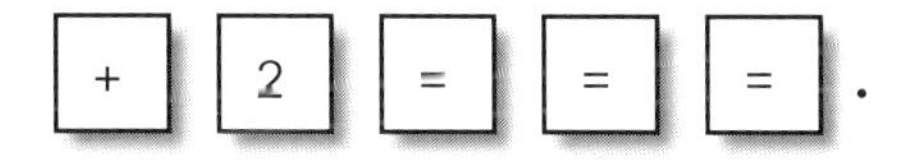

1	2	3	4	5	6	7	8	9	10
11	12	13	14	15	16	17	18	19	20
21	22	23	24	25	26	27	28	29	30
31	32	33	34	35	36	37	38	39	40
41	42	43	44	45	46	47	48	49	50
51	52	53	54	55	56	57	58	59	60
61	62	63	64	65	66	67	68	69	70
71	72	73	74	75	76	77	78	79	80
81	82	83	84	85	86	87	88	89	90
91	92	93	94	95	96	97	98	99	100

Recording the count on a number grid

Note: The display must be showing zero before entering this sequence, otherwise the calculator will continually add two to the number that was originally shown on the display.

Recording the result on a number grid will help to highlight patterns in the multiples of two, that is they all end in 0, 2, 4, 6, or 8. Students will begin to associate the multiples of two with the even numbers and begin to 'see' that even numbers end in 0, 2, 4, 6, or 8. Encourage and foster as many links as possible as this will help strengthen students' understanding of the 'two times table' and they are less likely to forget them.

Further work with 'doubling and halving' will help consolidate recall of the 'two times table'.

Note: While the 'three times table' is discussed next, if following a 'cluster approach' the two times table would be followed by the 'four and eight times tables'.

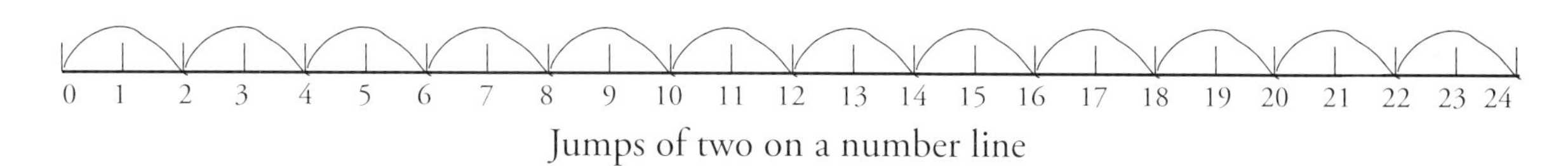

Jumps of two on a number line

Counting by two on a number track

x 3

Similar experiences to those outlined in the previous section on the 'two times table' may be undertaken to build familiarity with the 'three times table'. For example, a simple calculator may be set to count in threes by varying the keystroke sequence from:

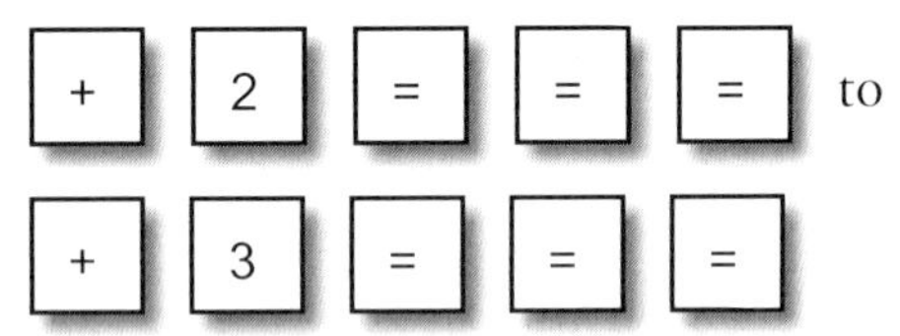

What is of particular interest with the 'three times table' is the pattern that is formed when the **digit sum** is calculated. **The *digit sum* or digital root as it is sometimes called involves adding all the individual digits that make up the number, until a single digit is formed.** For example, the digit sum of 328 is found by adding 3 and 2 and 8, which gives 13. This is still a two-digit number so you continue adding until a single digit answer is found. In this case, 1 + 3 = 4, so the digital root of 328 is 4.

An interesting pattern is formed when the digit sum of the 'three times table is examined. Ask students to list the multiples of three. Explain how to calculate the digit sum for a number and ask them to work out the digit sum for each of the multiples of three.

3; 3
6; 6
9; 9
12; 1 + 2 = 3
15; 1 + 5 = 6
18; 1 + 8 = 9
21; 2 + 1 = 3
24; 2 + 4 = 6
27; 2 + 7 = 9
30; 3 + 0 = 3

A close examination of the pattern reveals that the sequence 3, 6, 9 repeats. Students may investigate whether this pattern continues.

If students mark the multiples of three on a number grid then they can examine the digit sum pattern and consider the visual pattern that is created.

0	1	2	**3**	4	5	**6**	7	8	**9**
10	11	**12**	13	14	**15**	16	17	**18**	19
20	**21**	22	23	**24**	25	26	**27**	28	29
30	31	32	**33**	34	35	**36**	37	38	**39**
40	41	**42**	43	44	**45**	46	47	**48**	49
50	**51**	52	53	**54**	55	56	**57**	58	59
60	61	62	**63**	64	65	**66**	67	68	**69**
70	71	**72**	73	74	**75**	76	77	**78**	79
80	**81**	82	83	**84**	85	86	**87**	88	89
90	91	92	**93**	94	95	**96**	97	98	**99**

The visual pattern created on a number grid is quite pronounced. Students can 'see' the diagonal lines of multiples of three. There is considerable opportunity to make links between the 'three, six and nine times tables'. The cluster approach to learning the tables emphasises the links between the 'two, four and eight times table' and the 'three, six and nine times table'. Creating two more number grids showing the 'six and nine times tables' will help to establish those links. *A good grasp of doubling and halving is also required.*

Once it has been established that the digit sum for the three times table continues beyond thirty, students can investigate much larger numbers such a 4384 and 7287 to compare their digit sums and then divide them by three to find out whether they are multiples of three. Links may then be made to the rule for divisibility by three, that is, **if the digit sum is either 3, 6 or 9, then the number will be divisible by three without leaving a remainder.**

Note: While the 'four times table' is discussed next, if following a 'cluster approach' the 'three times table' would be followed by the 'six and nine times table'.

x 4

If following a cluster approach to teaching and learning the basic multiplication facts, then just prior to learning the 'four times table', the students would have learned the 'two times table'. ***As the 'four times table' will be built upon students' understanding of the 'two times table' they will have become fluent with them before embarking on learning the 'four times table'. As a doubling strategy will be employed in learning the 'four times table' students will need to have experience with the doubling strategy.***

The link between the 'two and the four times tables' may be established by colouring in the multiples of two and the multiples of four on a number grid.

1	2	3	4	5	6	7	8	9	10
11	12	13	14	15	16	17	18	19	20
21	22	23	24	25	26	27	28	29	30
31	32	33	34	35	36	37	38	39	40
41	42	43	44	45	46	47	48	49	50
51	52	53	54	55	56	57	58	59	60
61	62	63	64	65	66	67	68	69	70
71	72	73	74	75	76	77	78	79	80
81	82	83	84	85	86	87	88	89	90
91	92	93	94	95	96	97	98	99	100

Students will often make comments like "every second one is coloured twice." Encourage students to discuss this pattern and to use words and phrases like 'double' and 'multiplied by two'

A Venn diagram may be drawn to show the links between the multiples of two and four.

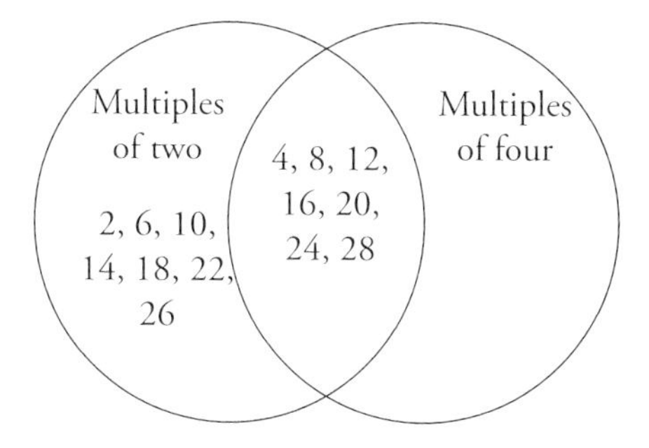

Drawing a Venn diagram helps the students to 'see' that all the multiples of four are also multiples of two. Discussion may then focus on how 'knowing the two times table' can help you 'work out' the 'four times table'.

Develop the Doubling Strategy

Review the doubling strategy for the 'two times table'. For example 8 x 2 is the same as double eight. Note: This requires making use of the commutative property of multiplication, that is 8 x 2 gives the same result as multiplying 2 by 8. A diagram often helps.

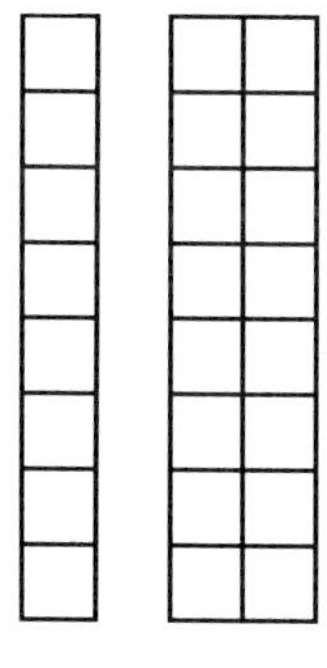

8 16

Then the diagram may be extended to 'double double' to illustrate the development of the 'four times table'. To work out 4 x 8 (or 8 x 4) simply start with eight and double and then double again.

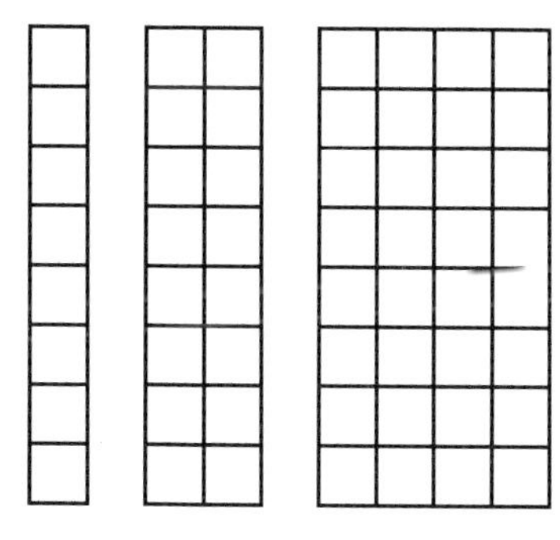

8 16 32

So 4 x 8 (2 x 2 x 1 x 8 or double 8, double 8) is 32. Students require experience with grid paper [See p. 33] in order to establish this thinking. ***Of course, eventually the students would be expected to develop fluency with the 'four times table'.***

x 5

Some teachers prefer to develop the 'five times table' before the 'three and four times tables'. Prior to working on the 'five times table', they make sure that the students understand and can apply the multiplication property of zero (any number multiplied by zero is zero) and the multiplication property of one (any number multiplied by one is itself). Next, they move on to the 'two times table' and then they often jump to the 'ten times table'.

While the 'ten times table' is not really part of the basic number facts, it does make sense as the 'ten times table' is simple to learn because of the pattern involved and may be used to support the learning of other basic multiplication facts. Remember, a powerful mental strategy involves using a known fact to derive an unknown fact. One such link may be drawn between the 'ten times table' and the 'five times table'. If a student can halve then multiplying by five is simple, just multiply by ten and halve the answer. As all multiples of ten end in zero it is a fairly simple matter to halve the result. For example, to work out 6 x 5, think 6 x 10 is 60 and halve to make 30. While I find this approach cumbersome, some students will find this approach quite novel and after applying it many times will develop fluency with the 'five times table'.

Another reason for following the sequence '0 x, 1 x, 2 x, 10 x table' is that students experience success at 'learning the tables' and develop a positive attitude toward learning them. Remember "nothing succeeds like success". Students can be given a blank tables grid and colour in all the table facts they know. They will soon realise they already know quite a few [See "Reducing the Memory Load", p. 12]. This helps weaker students especially see that the process of learning 'the tables' is not quite so daunting!

Consider the Pattern

Most students do not experience trouble learning the 'five times table'. The 5, 0, 5, 0 5, 0 pattern becomes most evident when counting by fives. Some students, however fail to notice this pattern. It may be emphasised in several ways.

A simple calculator may be set to 'count in fives' using the keystroke sequence:

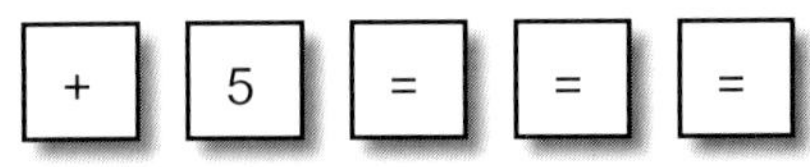

Sometimes students do not notice the five, zero pattern so ask them to cover part of the calculator display so that only the units digit is showing. The pattern will become much clearer. Colouring in a number grid will also help to emphasise the patterns. Students may notice that when five is multiplied by an odd number the answer always ends in five and when five is multiplied by an even number it ends in a zero, that is, it is even.

1	2	3	4	5	6	7	8	9	10
11	12	13	14	15	16	17	18	19	20
21	22	23	24	25	26	27	28	29	30
31	32	33	34	35	36	37	38	39	40
41	42	43	44	45	46	47	48	49	50
51	52	53	54	55	56	57	58	59	60
61	62	63	64	65	66	67	68	69	70
71	72	73	74	75	76	77	78	79	80
81	82	83	84	85	86	87	88	89	90
91	92	93	94	95	96	97	98	99	100

Links may be made with the divisibility rule for five, that is, **a number is divisible by five if it ends in a five or zero.** There are also obvious links to the clock as the interval between 12 and 1, 1 and 2 and so on is 5 minutes.

Students generally become fluent with the five times table quite quickly. The 'five times tables' then becomes a set of known facts from which other facts may be derived. For example, if you know 7 x 5 then via the commutative property you should also already know 5 x 7.

Because the 'five times table' is fairly easy to learn, there are plenty of opportunities to make use of it in developing strategies such as doubling and halving. For example, ask students to describe two ways of calculating 5 x 6, one that involves doubling and the other that uses halving.

x 6

Generally, by the time students reach the six times table they will have built up a bank of known facts that they can fluently recall. These will include the '0 x, 1 x, 2 x, 3 x 5 x and 10 x' (although the order may vary). If this is the case the students will already be fluent with the following facts: 0 x 6, 1 x 6, 2 x 6, 3 x 6, 5 x 6 and 10 x 6. If you know 2 x 6, you can work out 4 x 6 by doubling. Similar thinking may be applied to 3 x 6 and 6 x 6.

The students may not have learned the 'four times table', especially if a cluster approach, 3, 6, 9, 2, 4, 8 is being applied. Note: Some teachers may not tackle the six times table until the four and eight times tables have been learned.

Students will already have made use of strategies such as *doubling and halving* and *relate to a known fact* and will be aware of the properties of multiplication such as the *commutative property*. All of this knowledge may be applied to the teaching and learning of the 'six times table'.

Relate to a Known Fact

This is a very powerful strategy that relies on using a bank of known facts to derive other facts. For example, if you already 'know' 6 x 10, and therefore 10 x 6 (by applying the commutative property of multiplication) you may use this knowledge to 'work out' other facts such as 9 x 6 and 11 x 6. This strategy, 'relate to a known fact' is extremely powerful. Ask students to explain how they could use 10 x 6 = 60 to 'work out' what 9 x 6 is. Most students will subtract 6 from 60 to calculate the result of 54.

Similar reasoning may be used to 'work out' the product of 4 and 6 and 6 and 6. If students know that 5 x 6 is 30, they may apply this knowledge to work out that 4 x 6 is one lot of six less than 30 and 6 x 6 is one lot of 6 more than 30.

Doubling

The doubling strategy is at the heart of the cluster method of learning the tables, whereby the '3, 6 and 9 x tables' are linked. In this case it is assumed that students are already fluent with the 'three times table'. Links between the 'three' and the 'six times table' need to be established. One way to do this is to colour all the multiples of three and multiples of six on the same number grid. Some students will start to notice that every second cell is already coloured. Others will notice that the multiples of six form diagonals in much the same way as the multiples of three did, except they 'skip' a line (row).

0	1	2	3	4	5	**6**	7	8	9
10	11	**12**	13	14	15	16	17	**18**	19
20	21	22	23	**24**	25	26	27	28	29
30	31	32	33	34	35	**36**	37	38	39
40	41	**42**	43	44	45	46	47	**48**	49
50	51	52	53	**54**	55	56	57	58	59
60	61	62	63	64	65	**66**	67	68	69
70	71	**72**	73	74	75	76	77	**78**	79
80	81	82	83	**84**	85	86	87	88	89
90	91	92	93	94	95	**96**	97	98	99

Similar activities to those used to establish the 'three times table' may be used to develop the 'six times table'. For example, students may like to investigate whether the digit sum pattern they found in the 'three times table' works for the 'six times table'. The students will notice that the sum of the digits is either 3, 6 or 9. What they may not notice is that in each case the number is even. Discuss why this is the case and the link between doubling. Note: Doubling a number will always make the result even. The six times tables are 'double the three times table'. **The divisibility rule therefore is that the number is even and the digit sum is either 3, 6 or 9.**

The doubling strategy may be applied in different ways. For example, 3 x 6 may be thought of as 3 x 3, doubled. Alternatively to work out 3 x 6 you can double and add on the number. For example, 6 x 3 is double 6, which is 12 add on 6, which makes 18.

x 7

It makes sense to leave the seven times table until last. By this time these table facts will already have been developed when learning the other table facts. The 'eight times table' may be learned by relating them to the 'two and four times table'. There are many techniques for learning the 'nine times table' including using your fingers as a calculator.

Here are the 'seven times table' facts that the students will know already.

0 x 7 = 0 Multiplication property of zero.

1 x 7 = 7 Multiplication property of one.

2 x 7 = 14 'Two times table', application of the commutative property of multiplication, use of doubling strategy.

3 x 7 = 21 Relate to 'three times table' using the commutative property, application of the double and one more strategy.

4 x 7 = 28 Relate to 'four times table' via commutative property of multiplication, relate to the 'two times table' using a doubling strategy, application of the double-double strategy. Relate to 5 x 7, by thinking that 4 x 7 is one less 7 than 5 x 7.

5 x 7 = 35 Relate to the 'five times table' using the commutative property of multiplication, relate to the 'ten times table' using a halving strategy.

6 x 7 = 42 Relate to the 'six times table' via the commutative property of multiplication, relate to the 'three times table', 3 x 7, using a doubling strategy. Relate to 5 x 7, by thinking that 6 x 7 is one more 7 than 5 x 7.

7 x 7 = 49 An example of a square number. Square numbers may be treated as a separate group of 'table facts' to be learned.

8 x 7 = 56 If the students know the square number fact 7 x 7 then they can make use of the relate to a known fact strategy to work out the answer. Alternatively they could relate 8 x 7 to 2 x 7 or 4 x 7, using a doubling strategy. They could also choose to double 7, double again and double once more.

9 x 7 = 63 Use commutative property of multiplication to relate to 7 x 9. They may relate to 'ten times table' and think 10 x 7 – 7 or apply various patterns related to the 'nine times table'.

Eventually students need to memorise all the basic multiplication facts. This is certainly the case for the 'seven times table'. Students need to learn that while strategies may be used to learn the basic multiplication facts, they need to become fluent in recalling the basic multiplication facts.

Earlier reference was made to the use of calculators to support the learning of basic multiplication facts. The keystroke sequence

[+] [7] [=] [=] [=]

may be used to set a simple calculator counting by seven. The keystroke sequence

[2] [x] [7] [=] check order for each calculator

will generate the result of multiplying two by seven. Pressing [3] without clearing the display will produce a result of 21.

Pressing [6] will produce a result of 42 and so on.

x 8

The strategies previously discussed when dealing with the 'two times' and 'four times table' may be applied to the learning of the 'eight times table'. However, there is an added advantage at this point because the students will already have developed fluency with the 'two and four times tables'. ***Some students may not make the connection with these related 'tables' so they need to be made explicit.***

Colouring the multiples of four and eight will help students to spot the relationship between the 'four and eight times table'.

1	2	3	4	5	6	7	8	9	10
11	12	13	14	15	16	17	18	19	20
21	22	23	24	25	26	27	28	29	30
31	32	33	34	35	36	37	38	39	40
41	42	43	44	45	46	47	48	49	50
51	52	53	54	55	56	57	58	59	60
61	62	63	64	65	66	67	68	69	70
71	72	73	74	75	76	77	78	79	80
81	82	83	84	85	86	87	88	89	90
91	92	93	94	95	96	97	98	99	100

Most students will notice that when colouring in the multiples of eight, every fourth cell has already been coloured. Doubling the 'four times table' produces the 'eight times table'.

Another form of doubling strategy may also be used. For example, to work out 7 x 8 would involve doubling 7, doubling again and then double again.

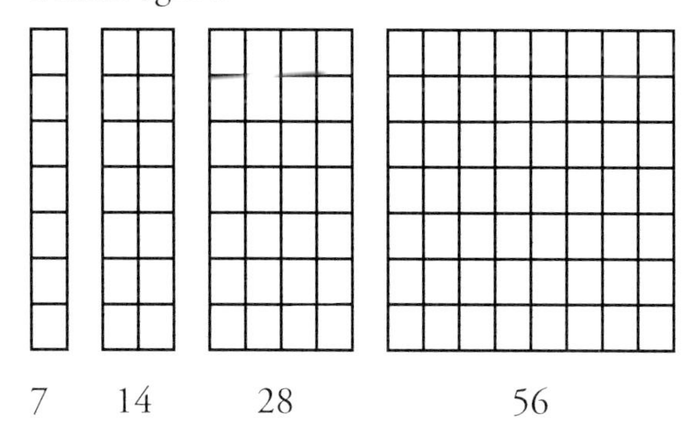

This strategy may be summarised as **double-double-double.**

Function machines (see below) can be used to introduce the double-double-double strategy. Explain that when a number enters a function machine it is changed before it comes out the other side. Six eights are forty-eight may be calculated using the double-double-double strategy as illustrated by the function machines below.

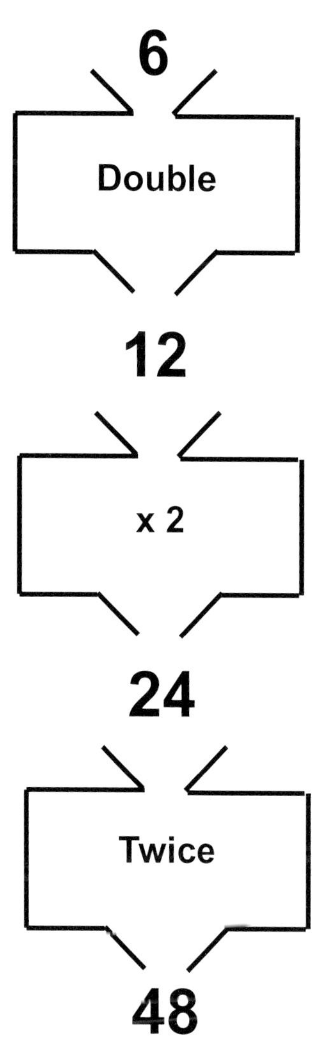

It should be noted that the the basic multiplication fact 7 x 8 = 56 and the equivalent 8 x 7 seem to cause the most difficulty for students and adults. (See page 60, Devlin, K. (2000) *The Mathematical Gene*. London: Weidenfeld & Nicholson)

x 9

The 'nine times table' is one of the simplest to learn as there are so many patterns and strategies that are associated with it. Because of this, many teachers choose to teach the nine times table before the 'seven and eight times tables'. Teachers who adopt the cluster approach to teaching the 'tables' will relate the 'nine times table' to the 'three and six times tables'.

Ask the students to list the multiples of nine and describe the patterns they notice.

1 x 9 = 9

2 x 9 = 18

3 x 9 = 27

4 x 9 = 36

5 x 9 = 45

6 x 9 = 54

7 x 9 = 63

8 x 9 = 72

9 x 9 = 81

Some of the patterns they will observe include:

- **The sum of the digits is always nine.**
- **The tens digit increases in order from 1 to 8.**
- **The units digit decreases by one each time from 9 to 1.**
- **The first answer is odd, the second even and so on.**
- **The tens digit is always one less than the non-nine factor. For example, the tens digit for 6 x 9 will be one less than the 6, that is 5 or 50. Given the sum of the digits must add to nine the units digit will be 4, hence the result is 54.**

A simple finger tables method may be used to multiply by nine. [See Finger Multiplication 1, p. 58]

Shading the multiples of nine on a number grid not only provides visual reference but also highlights the relationship to the 'ten times table'.

1	2	3	4	5	6	7	8	**9**	10
11	12	13	14	15	16	17	**18**	19	20
21	22	23	24	25	26	**27**	28	29	30
31	32	33	34	35	**36**	37	38	39	40
41	42	43	44	**45**	46	47	48	49	50
51	52	53	**54**	55	56	57	58	59	60
61	62	**63**	64	65	66	67	68	69	70
71	**72**	73	74	75	76	77	78	79	80
81	82	83	84	85	86	87	88	89	**90**
91	92	93	94	95	96	97	98	**99**	100

Ask the students to describe the pattern that they notice. Most students will refer to the diagonal that is formed running from the top right of the grid to the bottom left. Ask the students to explain why this pattern occurs. *Reference can be made to the "relate to a known fact strategy". Notice how the nine facts are related to the tens facts.*

1 x 9 = 1 x 10 – 1

2 x 9 = 2 x 10 – 2

3 x 9 = 3 x 10 – 3

4 x 9 = 4 x 10 – 4

5 x 9 = 5 x 10 – 5

6 x 9 = 6 x 10 – 6

7 x 9 = 7 x 10 – 7

8 x 9 = 8 x 10 – 8

9 x 9 = 9 x 10 – 9

This explains why the diagonal pattern is formed when the multiples of nine are shaded on a number grid.

Venn Diagrams

Purpose

To highlight the links between related table facts.

This activity supports the thinking behind the cluster method of learning the tables.

In this activity students are required to use Venn diagrams to sort and classify table facts according to whether they only belong to one specific group of tables, two groups or three. For example, in the Venn diagram containing the 'two, four and eight times tables 'the number 6 only belongs in the section belonging to the two times table and does not share a section with the four or eight times table. Of particular interest is that all the multiples of eight are also multiples of two and four. This is clearly shown in the Venn Diagram (above right).

In order to complete the Venn diagram, the students will need to list all of the multiples of two, four and eight and sort them according to which products are common to all three sets of table, only two sets of tables or just one.

Once the activity has been completed, the opportunity will arise to discuss the links between the 'two, four and eight times tables'. The doubling strategy that links the tables may be explored.

The same idea may be explored with the 'three, six and nine times tables'.

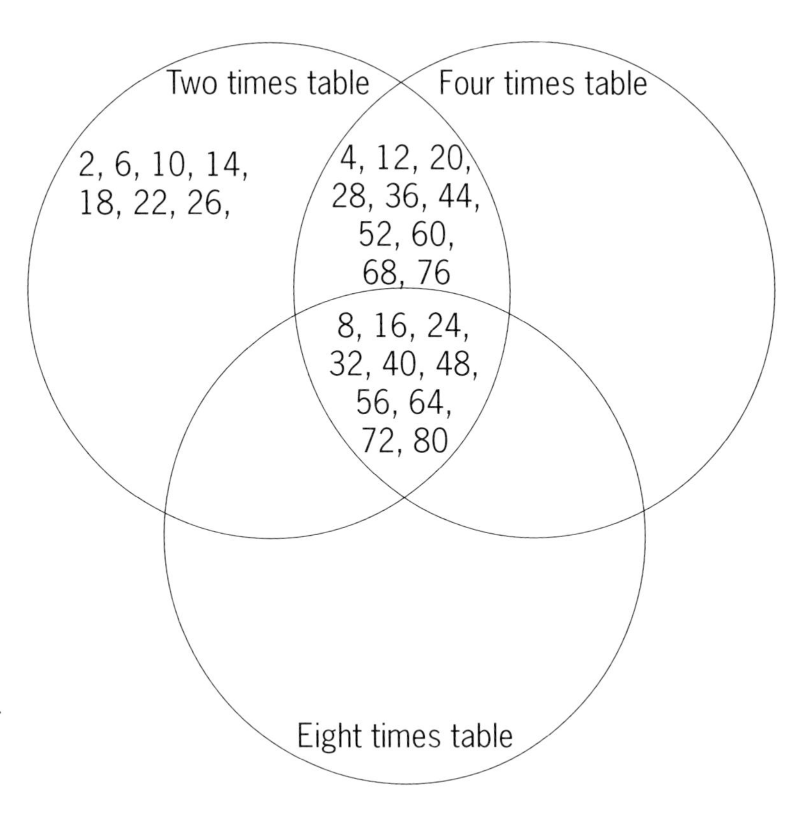

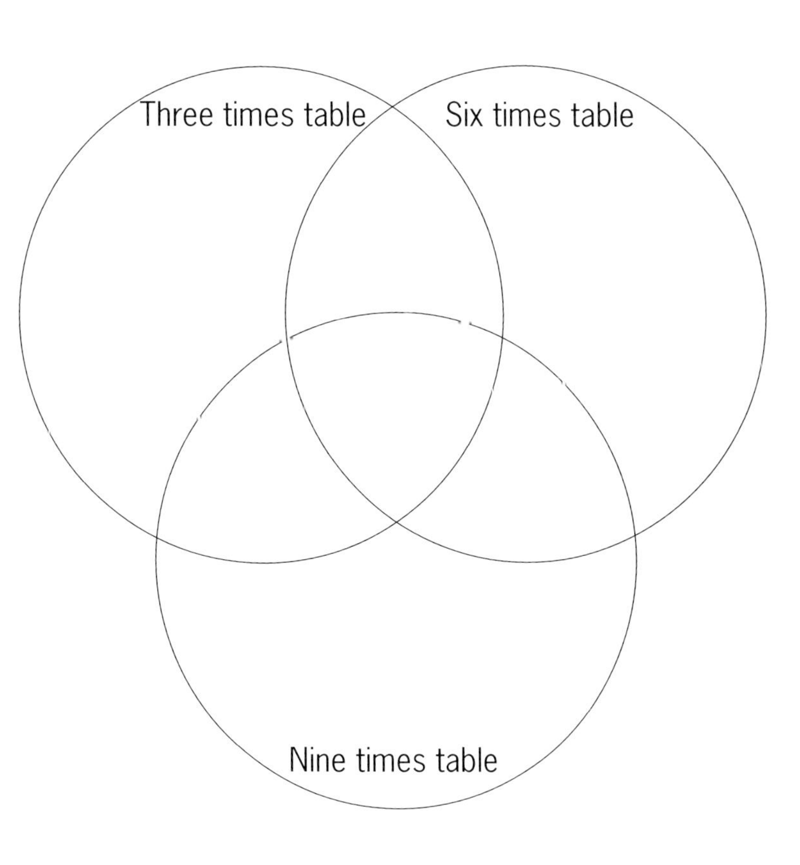

Square Numbers

After learning the 'zero, one, two, five and ten times tables' some teachers encourage their students to learn the square numbers. Often students find these fairly simple to learn as there is a verbal pattern that accompanies many of these multiplication facts. For example, 6 x 6 is 36 and 5 x 5 is 25 tend to 'roll off the tongue'.

While we refer to these numbers as 'square numbers' many students do not understand why they are called square numbers. Drawing grids with dimensions of 3 x 3, 4 x 4, 5 x 5 and so on with show why they are called square numbers – they form squares.

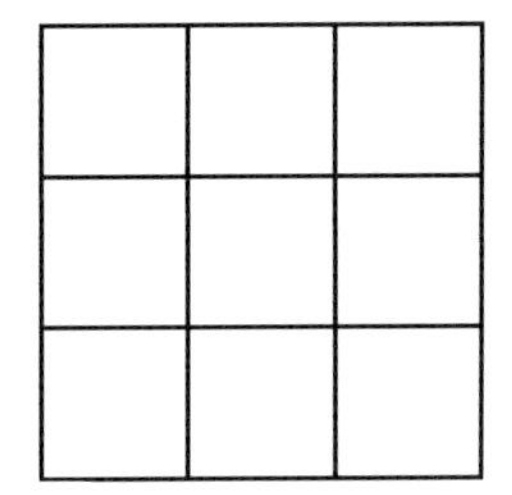

1	2	3	4	5	6	7	8	9	10
11	12	13	14	15	16	17	18	19	20
21	22	23	24	25	26	27	28	29	30
31	32	33	34	35	36	37	38	39	40
41	42	43	44	45	46	47	48	49	50
51	52	53	54	55	56	57	58	59	60
61	62	63	64	65	66	67	68	69	70
71	72	73	74	75	76	77	78	79	80
81	82	83	84	85	86	87	88	89	90
91	92	93	94	95	96	97	98	99	100

Locating the square numbers on a number grid shows that the gap between the square numbers increases. For example, the gap between 4 and 9 is 5, between 9 and 16 is 7, 16 and 25 is 9, 25 and 36 is 11 ... Notice how the pattern 5, 7, 9, 11 is formed.

x	1	2	3	4	5	6	7	8	9
1	1								
2		4							
3			9						
4				16					
5					25				
6						36			
7							49		
8								64	
9									81

Of more interest is to locate the square numbers on a multiplication grid. A diagonal is formed. Questions such as:

- What do you notice about the numbers directly above the square numbers? [They form the pattern, 0, 2, 6, 12, 20, 30, 42, 56, 72 and 90. The difference pattern is 2, 4, 6, 8, 10, 12, 14, 16, 18.]
- What do you notice about the numbers directly below the square numbers? [The pattern of numbers either side of the diagonal is the same. This highlights the symmetrical nature of the numbers either side of the square numbers, due to the commutative property of multiplication.]

The square numbers may be used to derive other basic multiplication facts. Using the 'relate to a known fact' strategy, students can 'work out' unknown basic multiplication facts. For example, 6 x 6 = 36 may be used to derive the fact 6 x 7. Adding one more 6 onto 36 will produce the result of 42. Likewise 7 x 7 = 49, may be used to 'work out' 7 x 8 is 56 – a notoriously difficult fact for some students (and adults) to remember.

Consolidation – Leading to Fluency

The activities that follow are designed to help students build on their basic fact knowledge. By this point students will have developed a bank of known facts and will continue to build upon these to increase their list of known facts. ***The activities contained in this section are designed to allow students enough time to 'work out' unknown facts using various strategies and multiplication properties.***

As the students gain familiarity with new facts and practise them without too much time pressure, they will increase their bank of known facts and develop fluency.

Triangular Flash Cards (p. 51)

Askew and Ebbutt refer to the use of 'free gifts' and triples when discussing the importance of making links and connections between multiplication facts. ***These Triangular Flash cards are ideal for developing connections between multiplication and division.*** Sets of cards may be made for various table facts. Remember, if children make their own, check that they are correct before allowing children to use them. As they employ self checking methods, children could practise getting the wrong answer. Variations of the Triangular Flash Card may be used to help make links between multiplication and division.

Fact Finder (p. 52 - 53)

This activity is designed to give children practice in standard tables questions (leftmost column) and related facts of the type 'number x ? = answer' and '? x number = answer'. ***These questions may be solved using division or by relating them to a known fact.*** Further questions may be produced with the aid of the template.

5 x 8 = 40, 8 x 9 = 72, 6 x 5 = 30, 3 x 6 = 18, 3 x 9 = 27, 4 x 7 = 28, 7 x 9 = 63,

4 x 8 = 32, 3 x 4 = 12, 5 x 7 = 35, 8 x 8 = 64, 6 x 6 = 36, 8 x 6 = 48, 3 x 3 = 9,

9 x 4 = 36, 6 x 5 = 30, 9 x 9 = 81, 3 x 7 = 21, 7 x 7 = 49, 7 x 2 = 14, 4 x 4 = 16, 8 x 7 = 56, 6 x 3 = 18

Finding Facts (p. 54 - 55)

The key to completing the grids is to look at those grid lines that contain two answers and then work backwards to find the original numbers on the outside of the grid. Further values may be found until the whole table is slowly built up.

Teachers and students can create their own puzzles by using the templates. Begin by filling in the outside of the grid. Next, complete the grid by multiplying. Then remove some numbers from the grid. Make sure enough of the grid remains so that the puzzle may be completed.

Multiplication Arithmagons (p. 56 - 57)

When choosing numbers to go into the squares, keep the result in mind. Multiplying the three numbers contained in the squares will produce a square number. The square root of this number is the same as the product of the numbers in the circles.

Starting from the left, moving to the top and then to the right of the Arithmagon, the answers are:

1) 9, 4, 5 2) 4, 2, 9 3) 8, 7, 3

4) 4, 8, 6 5) 5, 7, 6 6) 6, 4, 9

7) 7, 2, 9 8) 4, 7, 5 9) 8, 9, 7

10) 7, 4, 32 11) 6, 3, 21 12) 8, 72, 5.

Finger Multiplication 1 – 4 (p. 58 - 62)

Some teachers may be a little reluctant to introduce finger multiplication methods. ***Finger Multiplication 1 and 2 are particularly useful methods for older students who are not motivated to learn their tables.*** Older students (and adults) tend to like the quirky nature of finger multiplication methods. Finger Multiplication methods 3 & 4 have been included for interest.

Vedic Multiplication (p. 63)

This method is particularly useful for students experiencing difficulty learning the '7, 8 and 9 times table'. Vedic multiplication is often at the heart of many 'speed methods' for multiplying.

Vedic Square (p. 64 - 65)

Building a Vedic square involves completing all the basic multiplication facts from 1 x 1 to 9 x 9, giving the students valuable practice. Once the answers have been calculated the students then have to 'work out' the digit sum. **The digit sum, sometimes referred to as the 'digital root' is defined as the one-digit number produced when the digits in two-digit and larger numbers are added together. The digital root is used to examine patterns in the 'three and nine times table'.**

Last Digit Patterns (p. 66)

The following three activities all involve looking at the last digit in each set of basic facts. In order to complete this activity students need to list all the basic number facts. Focusing on the last digit helps students to notice patterns such as the 5, 0, 5, 0 pattern in the multiples of five.

The Ones Digit (p. 67)

Students might be surprised to find that the most likely digit(s) to appear in the units place are 2, 6 and 8. Looking for this answer involves listing all the basic multiplication facts and highlighting the last digit. Of the 81 basic number facts from 1 x 1 to 9 x 9, 25 are odd and 56 are even.

Circular Multiple Mazes (p. 68 - 69)

Once again this activity involves listing all the basic multiplication facts and then examining the last digit in each. The 'six and four times tables' produce the same patterns as do the 'eight and two-times tables' and the 'seven and three' and 'nine and one times tables'.

All The Answers (p. 70)

This is an extremely important activity as the focus is changed from calculating the product to determining the factors. Factorising is part of the algebra students will encounter in later years. Often we focus on calculating the product but this activity and the one that follows alter this thinking.

Product Pairs (p. 71)

Students are provided with the answers to basic multiplication facts and have to write the factors that produce the particular answer.

Marking Multiples (p. 72 - 73)

This activity helps to focus children on various patterns within the table facts. For example, the multiples of five only occur in two columns, the 5 and 10. Students may notice that all multiples of 5 end in five or zero. The geometric pattern formed by the multiples of nine illustrate the relationship between the nine and ten times table. i.e. 1 x 9 = 1 x 10 – 1, 2 x 9 = 2 x 10 – 2, 3 x 9 = 3 x 10 – 3, etc. In addition, when shading in the chart, children use a variety of strategies such as skip counting.

When the nine-times table is shaded on the 1 – 80 chart the diagonal will run from the top left to the bottom right. This occurs because the nine-times table is related to the eight - times table. i.e. 1 x 9 = 1 x 8 + 1, 2 x 9 = 2 x 8 + 2, 3 x 9 = 3 x 8 + 3, etc.

The 'five, three, eight and four times tables' will all run in a diagonal from the top right of the chart to the bottom left because all the charts are one column wider than the table facts being shaded.

The 'six and seven times tables' will run in a diagonal from the top left of the chart to the bottom right because these charts are all one column less than the table fact under consideration. Each table is related to the table one above or below it.

Multiplication Jigsaws (p. 74 - 76)

The patterns within the multiplication grid will assist students to re-form the grid once it has been cut up.

Flash Cards

Flash cards are often associated with boring drill activities, but the following variations on the humble flash card offer some useful alternatives.

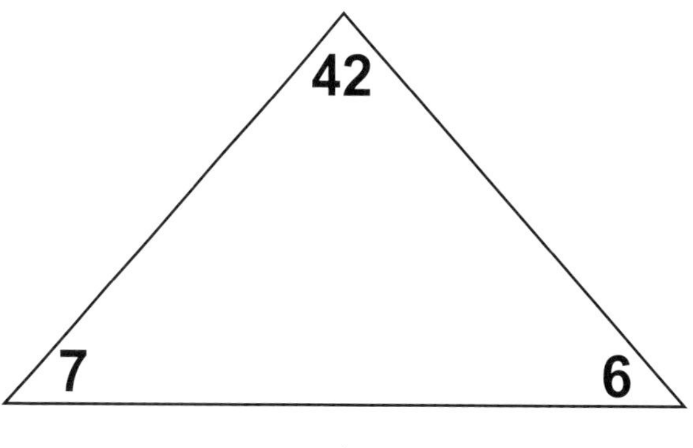

Triangular Flash Cards

Triangular flash cards are ideal for developing connections between related number facts. Often children learn one fact, such as 7 x 6 = 42, but cannot relate it to other facts, 6 x 7, 42 ÷ 7 and 42 ÷ 6 = 7.

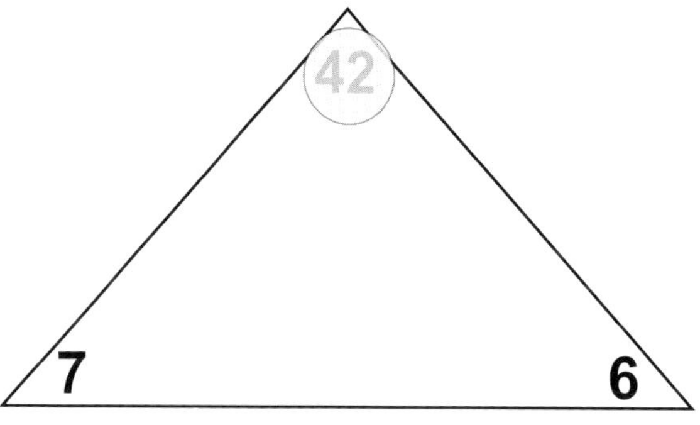

Triangular fact cards are designed to help children link various facts together. They are simple to make as illustrated .

The result is written at the top of the triangle and the factors at the base.

Students place a finger or thumb on one corner and are encouraged to state/write:
7 x 6 = 42, 6 x 7 = 42, 42 ÷ 6 = 7, 42 ÷ 7 = 6

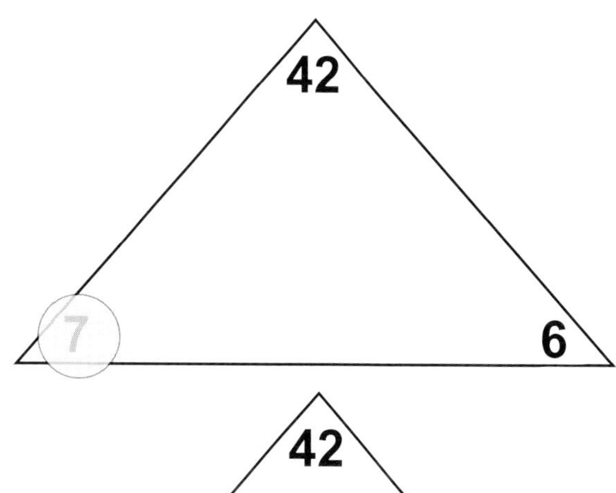

More able students can also add $\frac{1}{7}$ of 42 = 6, $\frac{1}{6}$ of 42 = 7

Children may work on their own, or in pairs placing their thumb over one corner of the card.

The following derivations of the triangular flash card work in a similar way.

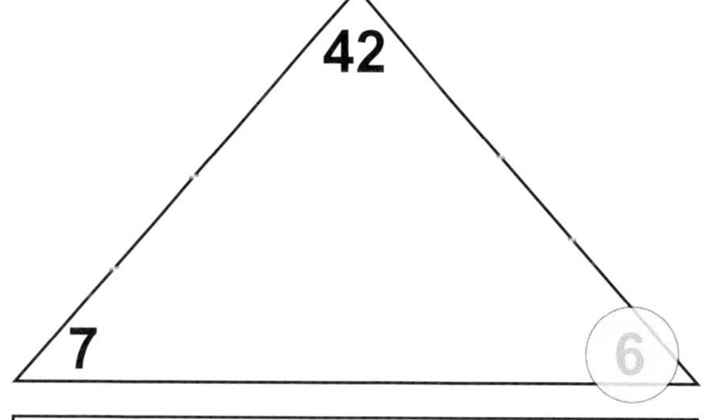

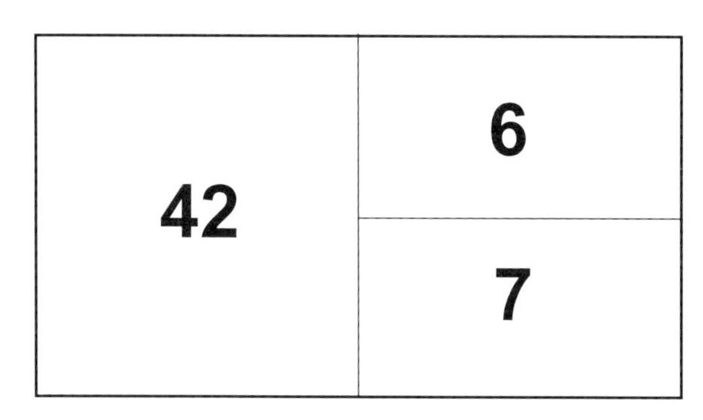

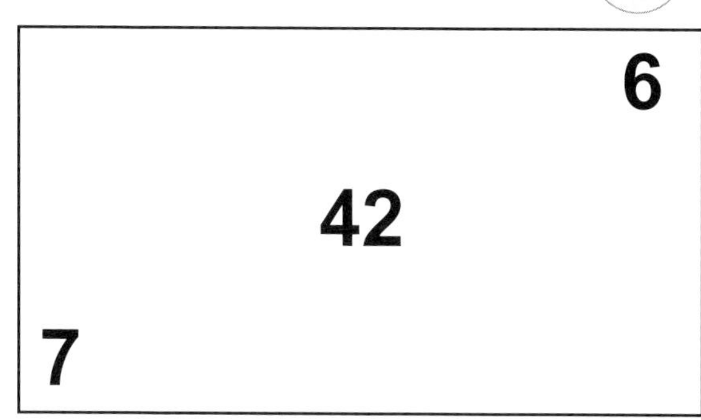

Fact Finder

The numbers in the boxes are related. For example, 7 x 6 = 42, 6 x 7 = 42, 42 ÷ 7 = 6 and 42 ÷ 6 = 7.

Use the relationships between the two numbers shown to 'work out' the missing third number.

7	6	42	5		40		9	72
6	5		3		18		9	27
4	7		7		63		8	32
3	4		5		35		8	64

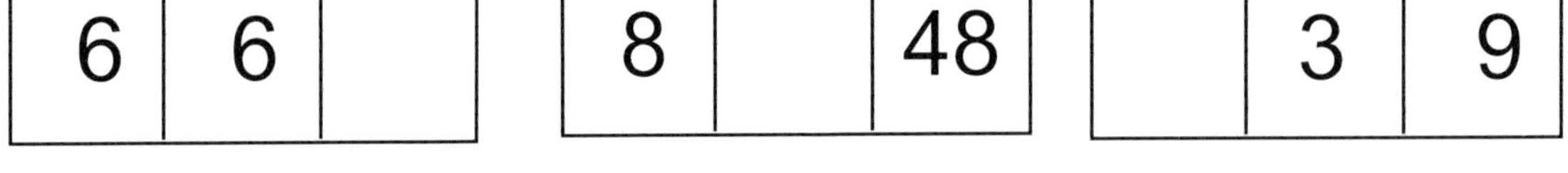

6	6		8		48		3	9

9	4		6		30		9	81
3	7		7		49		2	14
4	4		8		56		3	18

Fact Finder Blanks

The numbers in the boxes are related. For example, 7 x 6 = 42, 6 x 7 = 42, 42 ÷ 7 = 6 and 42 ÷ 6 = 7.

Try making some Fact Finder puzzles for a friend to solve.

7		42

Finding Facts

Use the numbers shown to work out the missing facts. The tables are not in order.

(1)

x	7			
4	28	36		
		18	6	
			24	48
				30

(2)

x		6		
	21	42		
11			55	44
				32
	30			

(3)

x	2		3	
	16			
4		40		
			21	42
				54

(4)

x				9
		15	40	
				81
4		12		
	6			54

The last two are very tricky. Look along the rows or down the columns for multiples of a specific fact.

(5)

x				
	21		12	
	49	14		
			36	45
			16	

(6)

x				
	35		14	
		60		48
		40	8	
	45			

Finding Facts

Make some missing fact problems of your own.

Begin by making up an entire grid, and then remove some numbers for your partner to work out. Remember to leave enough clues so the numbers can be worked out.

(1)

X				

(2)

X				

(3)

X				

(4)

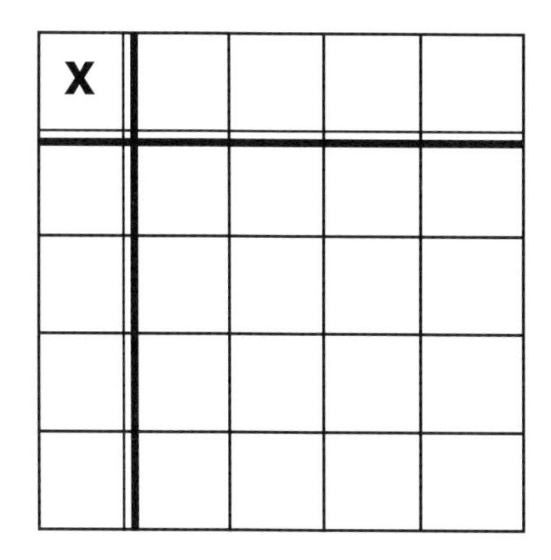

X				

(5)

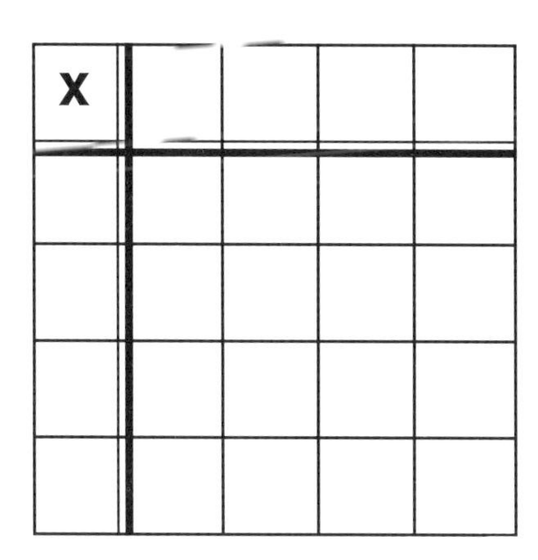

X				

(6)

X				

Multiplication Arithmagons

The numbers in the circles are multiplied to make the numbers in the squares between them. For example $3 \times 4 = 12$

$3 \times 9 = 27$

$4 \times 9 = 36$

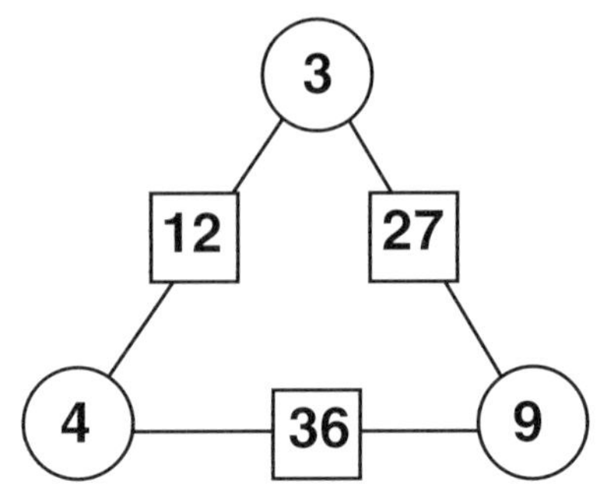

Work out the missing numbers.

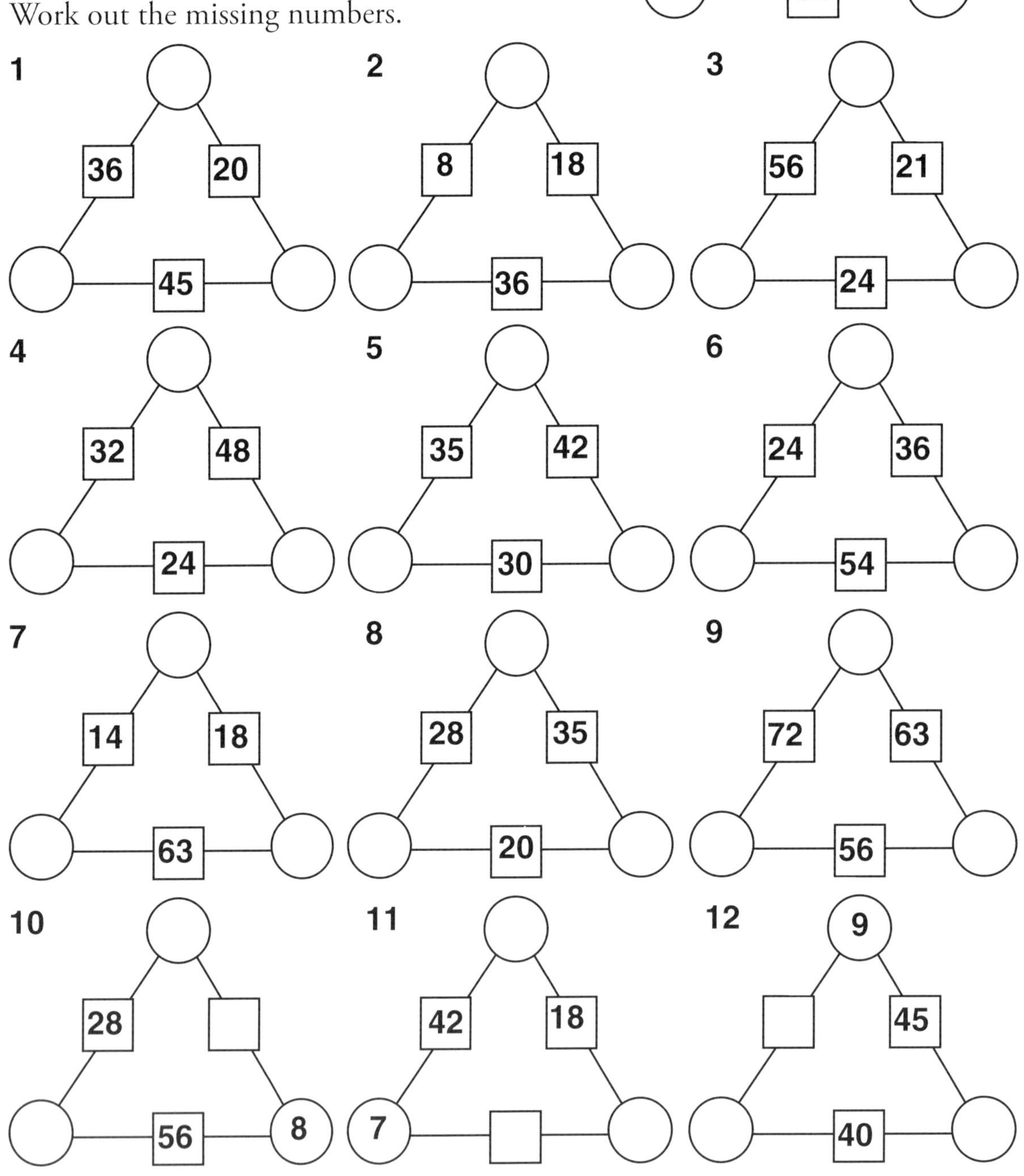

Try multiplying the numbers contained in the three squares of each arithmagon. Find the square root of the result. Now multiply the numbers in the circles. What do you notice?

Multiplication Arithmagons

The numbers in the circles are multiplied to make the numbers in the squares between them. For example 3 x 4 = 12

3 x 9 = 27

4 x 9 = 36

Make up some for a friend to solve.

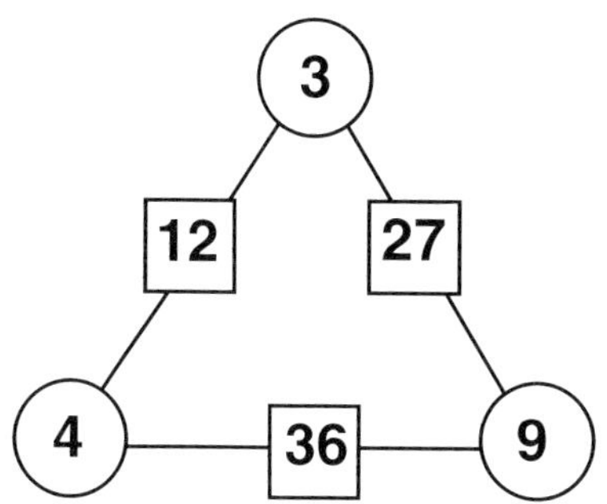

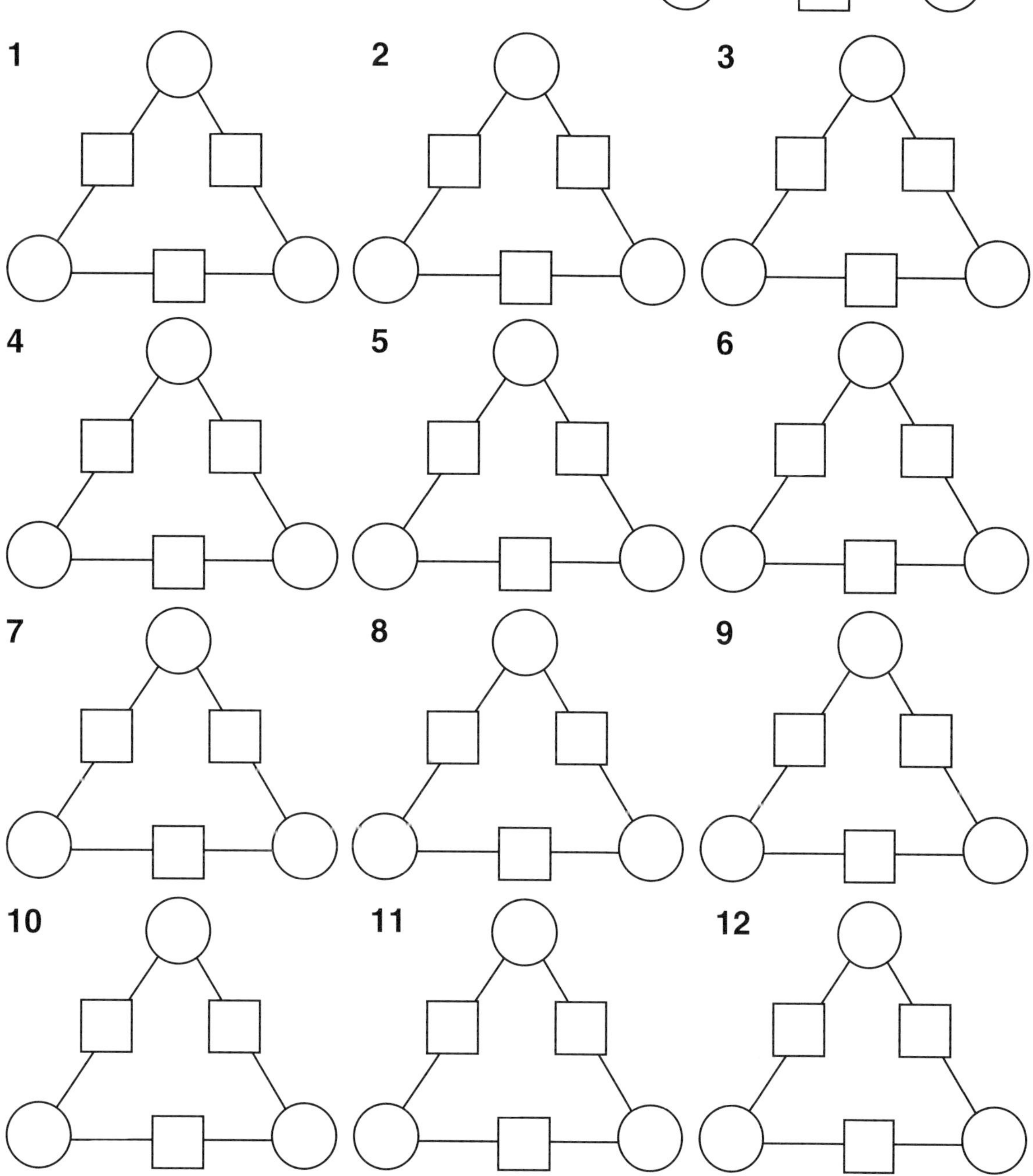

Finger Multiplication 1

This finger multiplication method works for the nine times table.

Ask pupils to hold hands in front of them and number them from 1-10 as shown:

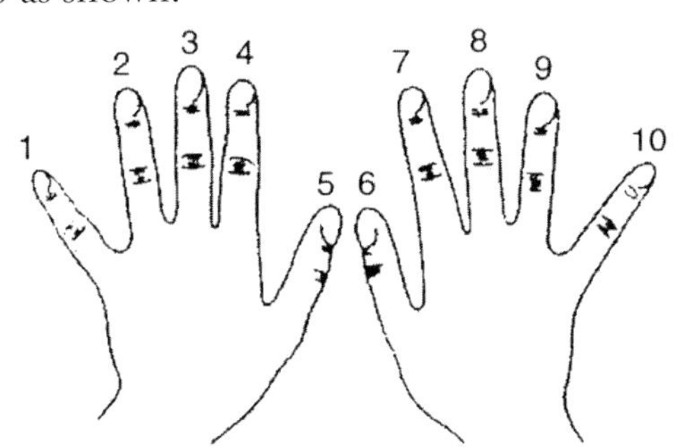

To multiply 3 x 9, bend down finger number 3. That leaves 2 fingers to the left of it and 7 to the right.

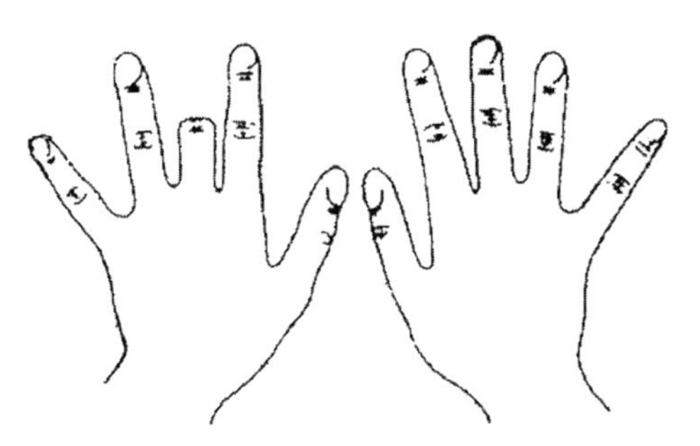

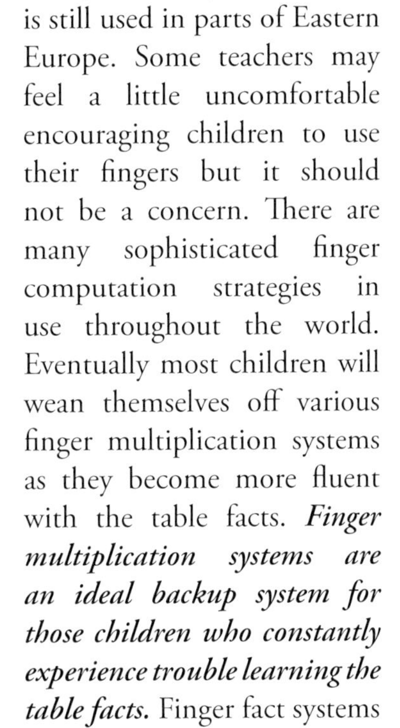

Peasant arithmetic or finger calculation has been used in Europe for many centuries and is still used in parts of Eastern Europe. Some teachers may feel a little uncomfortable encouraging children to use their fingers but it should not be a concern. There are many sophisticated finger computation strategies in use throughout the world. Eventually most children will wean themselves off various finger multiplication systems as they become more fluent with the table facts. ***Finger multiplication systems are an ideal backup system for those children who constantly experience trouble learning the table facts.*** Finger fact systems may also suit those children who have yet to master tables by the end of primary school.

The fingers to the left of the folded finger stand for the tens and the fingers to the right of it each stand for the ones, so 3 x 9 = 27.

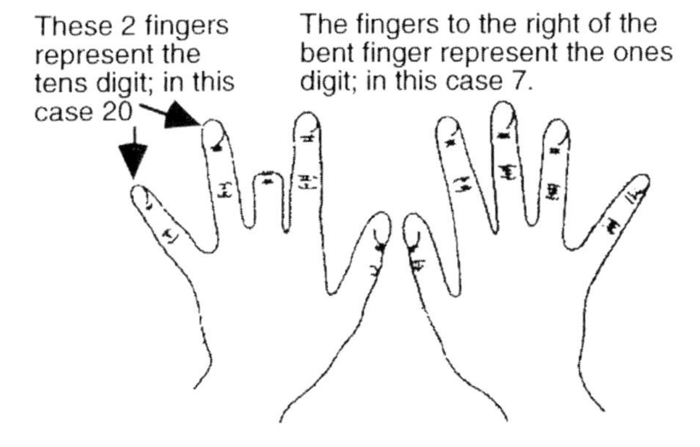

Variation

The finger multiplication method for the nine times table may be extended to include 12 x 9, 13 x 9, 14 x 9, 15 x 9, 16 x 9, 17 x 9, 18 x 9, 19 x 9 and 20 x 9. Unfortunately 11 x 9 doesn't work but this is easy to do in your head anyway.

For example to multiply 14 x 9.

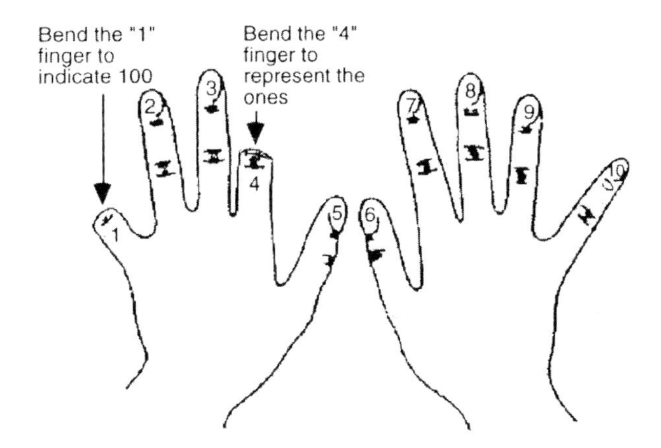

Bending the first finger is done automatically to indicate that the answer will be 1** (one hundred and something). The finger that corresponds to the units digit is then bent over. Count the number of fingers between the first bent finger and the second bent finger to find the tens digit. The ones digit is found by counting the remaining fingers.

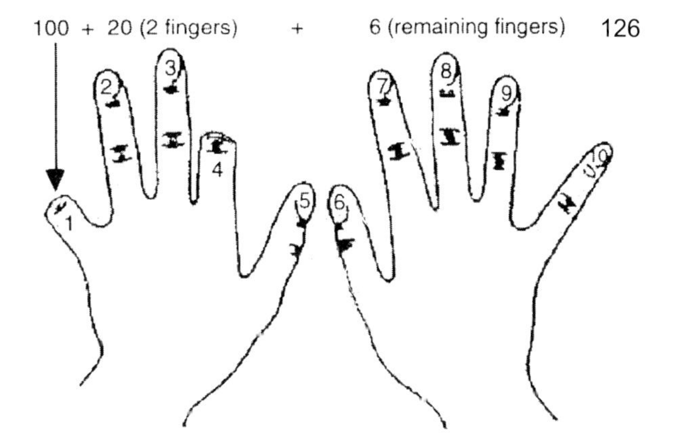

This same method may be used to multiply larger two-digit numbers by nine, providing the units digit is greater than the tens digit. For example, you could multiply 56 x 9 using this method, but not 54 because the tens digit is bigger than the units digit. In this case you would need to fold down five fingers to indicate the answer would be five hundred and something.

Finger Multiplication 2

Hold your hands with palms facing you and imagine each finger is labelled from 6 to 10, starting with your little pinkie. [See diagram below.]. To multiply seven by eight, touch the finger labelled 7 and the finger labelled 8 together to form a bridge.

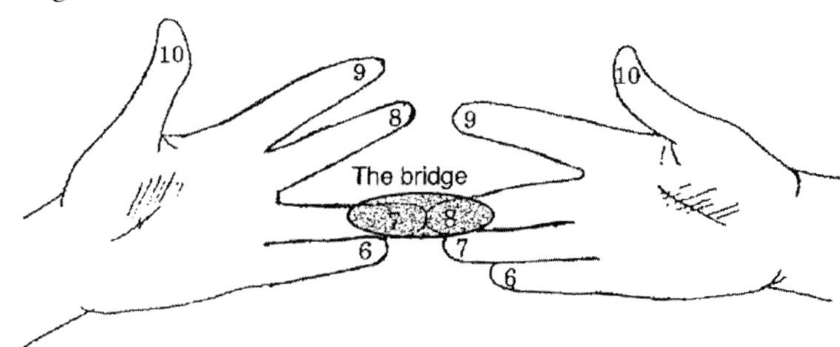

The number of tens in your answer is found by adding the number of fingers touching and below on one hand, to the number of fingers touching and below on the other hand.

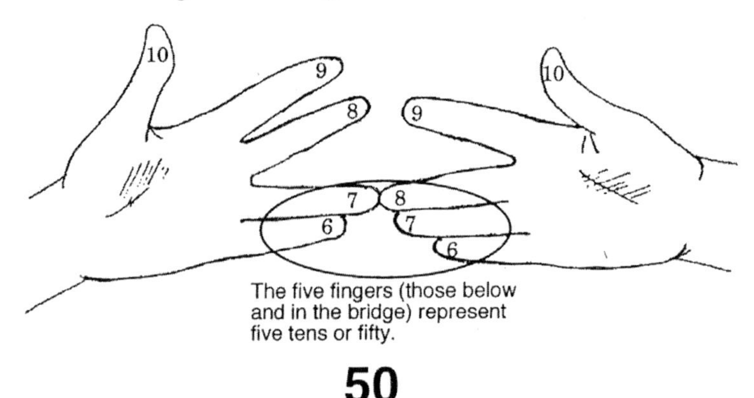

50

The number of ones is found by multiplying the number of fingers above the touching fingers.

6

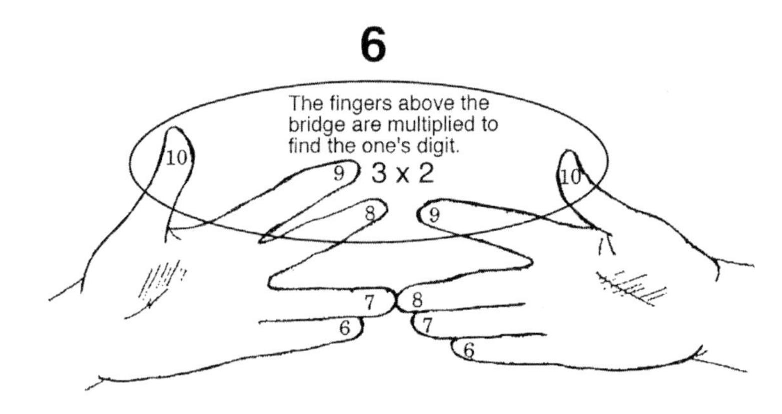

The answer to 7 x 8 is 56.

This finger calculation method works palms up or down. If working with the palms facing down, then the numbering would start at the thumb, that is, the thumb would be numbered 6 and the little pinkie would be 10. The instructions and diagrams have been given for palms facing up.

The finger tables method is suitable for tables from 6 x 6 to 10 x 10. It should be noted that when multiplying 6 x 6 and 6 x 7 using this method an extra step is required.

For example,

6 x 6 gives 20 + 16 = 36

6 x 7 gives 30 + 12 = 42

Finger Multiplication 3

The fingers on each hand should be labelled as shown.

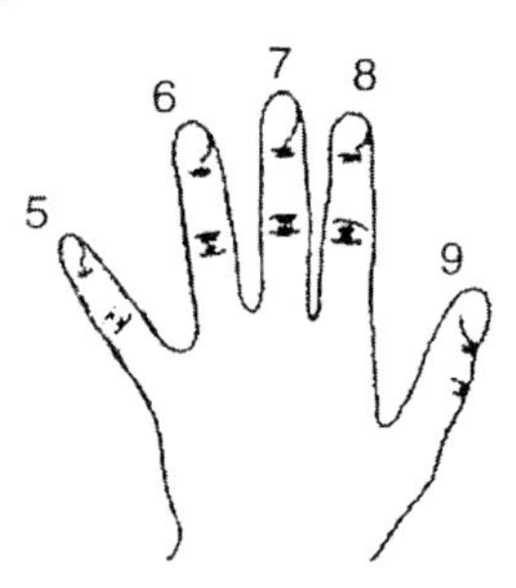

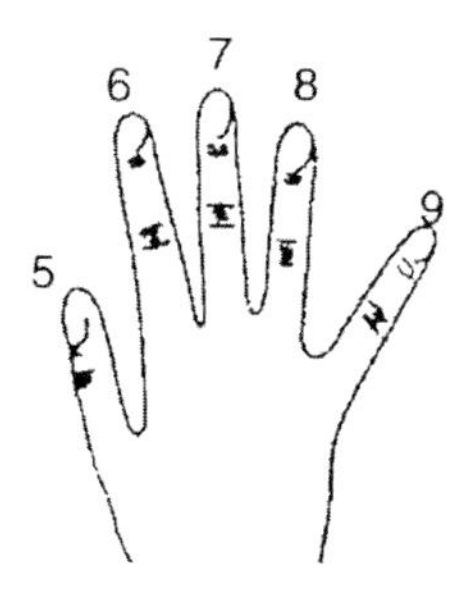

To work out 6 x 8 you would bend the finger labelled 6 on your left hand and the number 8 finger on the right hand.

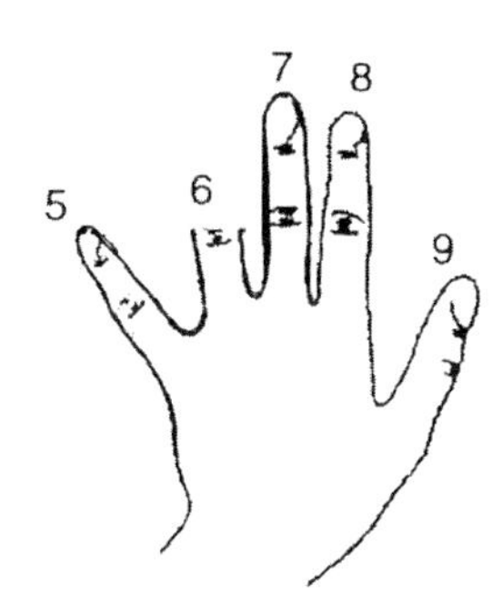

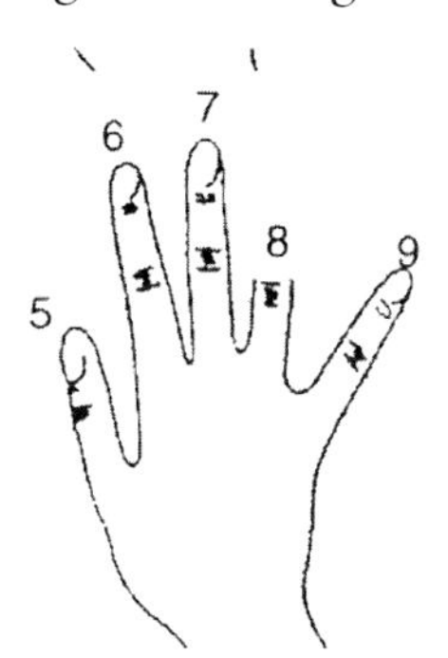

Note: While this method works for 5 x 5, 5 x 6, 6 x 6 and 6 x 7, it is somewhat clumsy to use and probably should be reserved for use with the higher tables.

Some teachers are a little hesitant to show finger multiplication to their students. it should be noted that finger arithmetic was commonplace hundreds of years ago.

Leonardo of Pisa (otherwise known as Fibonacci) in his book *Liber Abaci* which outlined the use of Hindu - Arabic numbers and various written algorithms, made the following comment:

"multiplication with the fingers must be practised constantly so that the mind, like the hands, becomes more adept at adding and multiplying". (as cited in Butterworth, B (1999) *The Mathematical Brain*. London, England: Macmillan p. 226).

Peasants in central France have been recorded using finger multiplication methods similar to those described in this book as late as the 1930's.

Next you count the fingers to the left of your bent finger on the left hand and the fingers to the left of your bent finger on your right hand. The **total of these fingers represents the tens' digit of the answer.**

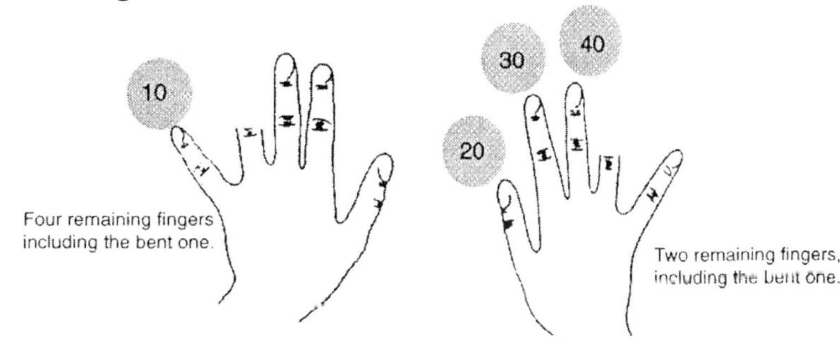

To find the ones' digit **multiply** the remaining fingers on the left hand by the remaining fingers on your right hand that is, 4 x 2). **Remember to include the bent fingers.** The answer to 6 x 8 is 48.

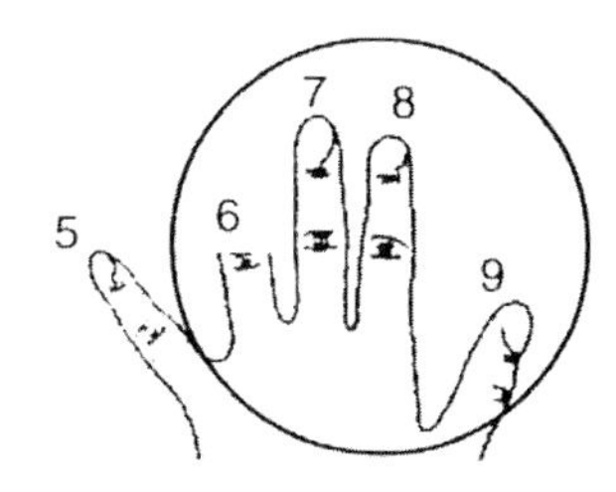

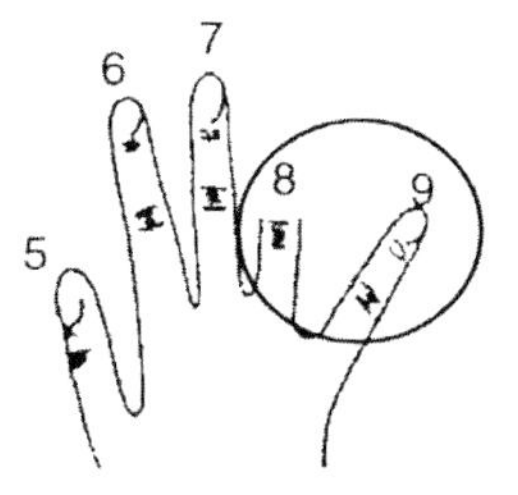

Finger Multiplication 4

Note: While this method may be used to multiply any two numbers between 5 and 10, it is a little tricky working with 5 x 5, 5 x 6, 6 x 6 and 6 x 7 and probably it should only be used with other tables.

Unlike the previous Finger Multiplication methods, this method begins with fingers (including the thumb) made into a fist. Fingers are *raised* rather than folded when performing a calculation.

For example, to calculate 6 x 8, follow these steps:

Determine the difference between the first number and five and raise that many fingers on your left hand. In the example above one finger would be raised because 6 – 5 = 1.

Determine the difference between the second number and five and raise that many fingers on your right hand. In this case 8 – 5 = 3, so three fingers are raised.

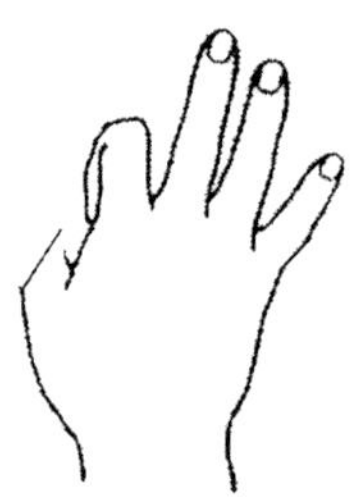

The tens digit is found by adding the raised fingers. In the above case, 1 + 3 = 4 tens or 40.

The units digit is found by counting the folded fingers (including the thumbs) on each hand and then multiplying them together. In this case 4 x 2 = 8.

The total is 48.

This Finger Multiplication method may be extended to include numbers greater than ten. For example, to multiply 12 x 13, two fingers would be raised on the left hand to indicate the difference between ten (not five as in the previous case) and the number to be multiplied. Likewise, three fingers would be raised on the right hand to show the difference between ten and thirteen.

- Add the raised fingers
  2 + 3 = 5 or 5 tens or 50.
- Multiply the folded fingers 3 x 2 = 6.
- Add the results
  50 + 6 = 56.
- Add 100.
  100 + 56 = 156.

How might this method be extended beyond 15 x 15?

Vedic Multiplication: Vertically & Crosswise

Children experiencing difficulty with the seven, eight and nine times table may like to try the "vedic method".

- Draw an 'X' on the page and write the numbers to be multiplied on the left. For example to multiply 7 x 8 you would write:

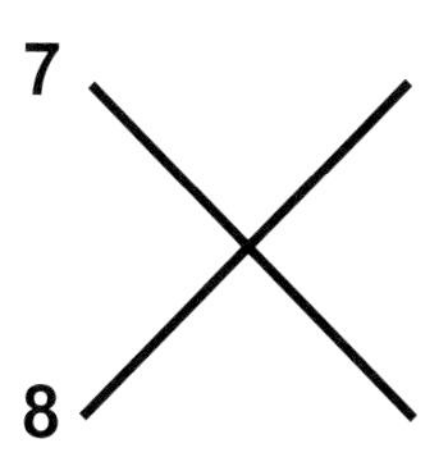

- Next subtract the top number (in this case 7) from 10 and write the result at the top right-hand end of the cross.

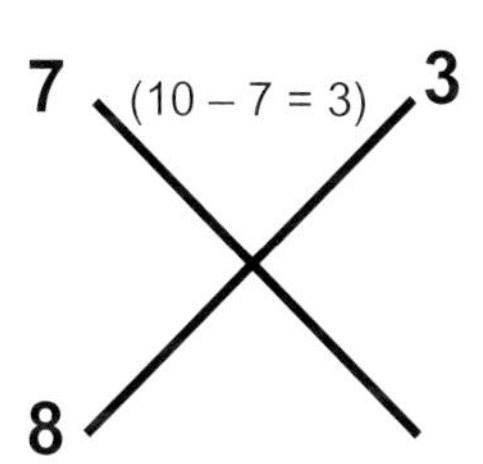

- Then subtract the bottom number (in this case 8) from 10 and write the result in the bottom right-hand corner of the cross.

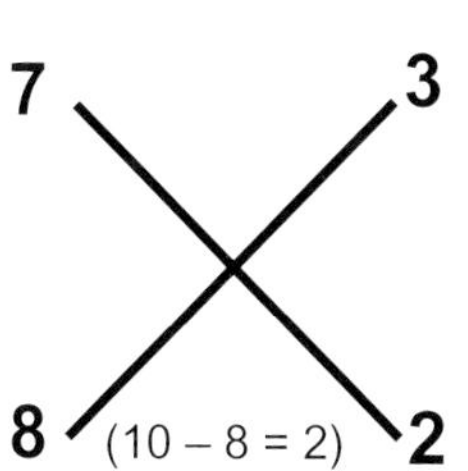

- To find the *tens digit* of the answer find the difference between the numbers at either end of the cross. It does not matter which pair you choose. For example, 7 – 2 gives five as the tens digit, that is 50.

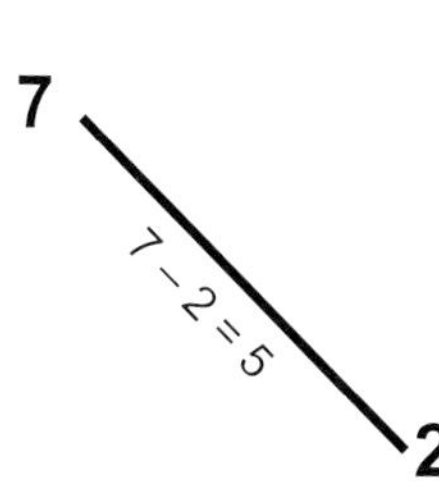

OR

- 8 - 3 also gives the tens digit so you know the answer will be fifty something (5?).

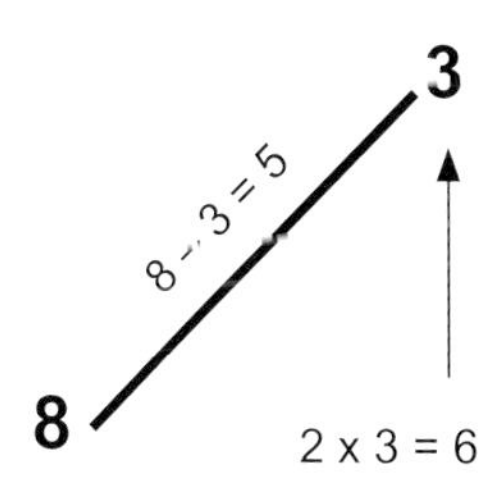

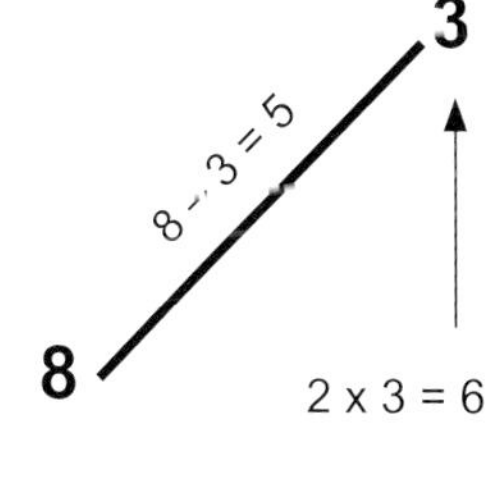

- To work out the ones digit simply multiply the two right hand digits of the cross. That is 2 x 3.

- Therefore the answer to 7 x 8 is 56.

This method may be slightly more complicated (it involves both multiplication and subtraction) and certainly less efficient than automatic recall of the 'seven eight and nine times tables' but it has the advantage that no numbers larger than 3 x 3 will need to be multiplied.

Background

Vedic maths comes from the Vedic tradition of India. The system was lost over the centuries and then rediscovered from ancient Sanskrit texts. The system is based on Sutras which describe ways of solving various problems. The Sutra "vertically and crosswise" may help children who experience trouble with table facts above 5 x 5.

The cross tables method relies on being able to subtract small numbers using ten as the reference point. This method is particularly useful for the 'seven, eight and nine times table' as it becomes a little messy with other numbers unless a few adjustments are made. For those children experiencing difficulties with the 'seven, eight and nine times table' (which represents a large group of students) this method simply provides another strategy by which the answer may be found. With a little practice, children can become quite adept at using this method. Eventually it is hoped that they would abandon this method in favour of automatic recall.

Note: This method may also be applied to other multiplication problems, for example, 96 x 97, using 100 as the reference point.

96 4

97 3

96 – 3 = 93 and 3 x 4 = 12, therefore the answer is 9312.

A word of caution. If children do not understand why this method works, then they will experience difficulties using this method with items like 97 x 98. Following the method would produce '95' as the first two digits and 3 x 2 produces 6 as the last digit. Clearly 956 is not the correct result, but 9506 is.

The Vedic Square

Background to Vedic Square Activity

[See the activity on page 65. These notes accompany the activity.]

This activity is thought to have originated in the Vedas (ancient Indian texts), hence the name 'Vedic Square'. Several patterns may be seen in the square. For example, the ninth row and column consist entirely of nines. If we forget the nines the first row is the reverse of the eighth row, the second is the reverse of the seventh, the third the reverse of the sixth and the fourth is the reverse of the fifth. The third and sixth rows consist of the same multiples of three in different order repeated three times.

There are various symmetrical patterns that may also be found. For example if you draw squares of increasing dimensions from the centre out you will notice bands of repeated digits. This is caused in part due to the symmetrical nature of the tables grid around the square numbers.

Students can explore other patterns such as what happens when all the ones, twos or threes are shaded?

If students experience difficulty creating the table to the right in a single step use the following two step procedure.

Ask the students to complete a standard 9 x 9 multiplication table. This will provide practice in recalling the basic multiplication facts.

Once the multiplication table has been completed then it may be used as a reference for completing the Vedic Square.

x	1	2	3	4	5	6	7	8	9
1	1	2	3	4	5	6	7	8	9
2	2	4	6	8	1	3	5	7	9
3	3	6	9	3	6	9	3	6	9
4	4	8	3	7	2	6	1	5	9
5	5	1	6	2	7	3	8	4	9
6	6	3	9	6	3	9	6	3	9
7	7	5	3	1	8	6	4	2	9
8	8	7	6	5	4	3	2	1	9
9	9	9	9	9	9	9	9	9	9

x	1	2	3	4	5	6	7	8	9
1	1	2	3	4	5	6	7	8	9
2	2	4	6	8	1	3	5	7	9
3	3	6	9	3	6	9	3	6	9
4	4	8	3	7	2	6	1	5	9
5	5	1	6	2	7	3	8	4	9
6	6	3	9	6	3	9	6	3	9
7	7	5	3	1	8	6	4	2	9
8	8	7	6	5	4	3	2	1	9
9	9	9	9	9	9	9	9	9	9

x	1	2	3	4	5	6	7	8	9
1									
2									
3									
4									
5									
6									
7									
8									
9									

The Vedic Square

Interesting patterns may be observed when the 'digit sums' of the table facts are placed in a chart. A digit sum is found by adding the digits of the answer to a table fact until a single digit is left. For example 3 x 8 = 24. The digit sum of the answer, 24, is 6 (i.e. 2 + 4). The digit sum for 7 x 8 involves two steps 7 x 8 = 56, 5 + 6 = 11, 1 + 1 = 2.

Complete the following 'digit sum' chart.

Some have been completed for you.

What pattern do you notice in the 9 x table?

What other patterns do you notice?

x	**1**	**2**	**3**	**4**	**5**	**6**	**7**	**8**	**9**
1	1	2	3	4	5	6	7	8	9
2	2								
3	3							24 **6** 2 + 4	
4	4			16 **7** 1 + 6					
5	5								
6	6				30 **3** 3 + 0				
7	7							56 **2** 5 + 6 = 11 1 + 1 = 2	
8	8		24 **6** 2 + 4				56 **2** 5 + 6 = 11 1 + 1 = 2		
9	9								

Last Digit Patterns

If we examine the **last digits** in a particular table, patterns may be noted. For example, the six times table forms the following pattern.

0, 6, 12, 18, 24, 30, 36, 42, 48, 54, 60

If we separate the last digit from the other digits the pattern becomes clearer.

0, 6, 2, 8, 4, 0, 6, 2, 8, 4, 0 Note how the pattern repeats.

Complete the following 'last digit' chart. The 6 x table has been completed for you.

x	0	1	2	3	4	5	6	7	8	9	10
0							0				
1							6				
2							2				
3							8				
4							4				
5							0				
6	0	6	2	8	2	0	6	2	8	4	0
7							2				
8							8				
9							4				
10							0				

Write about any patterns you notice.

Do the patterns continue?

What about the 'teens table'? For example, 13, 14, 15.

What happens with the 12, 23, 33 . . . times tables?

The Ones Digit

If two single-digit numbers (1 – 9) are multiplied at random, what digit is most likely to appear in the ones place?

If two single-digit numbers are multiplied at random, what are the chances that the result is even? What are the chances that the result is odd?

Make a guess and write it down.

Now answer the question complete the following tables chart.

x	1	2	3	4	5	6	7	8	9
1						6			
2						12			
3						18			
4						24			
5						30			
6						36			
7						42			
8						48			
9						54			

Keep a tally of the number of times each digit comes up in the ones place.

Ones digit	0	1	2	3	4	5	6	7	8	9
Tally										
Total										

Circular Multiple Mazes

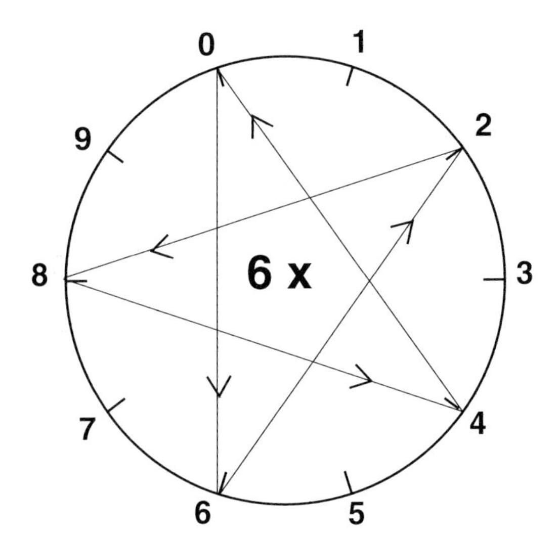

If we examine the last digits in a particular table, e.g. the six times table a pattern is formed.

0, 6, 12, 18, 24, 30, 36, 42, 48, 54, 60

becomes:

0, 6, 2, 8, 4, 0, 6, 2, 8, 4, 0

Note how the pattern repeats.

Some interesting geometric patterns may be found if the points are plotted on a ten

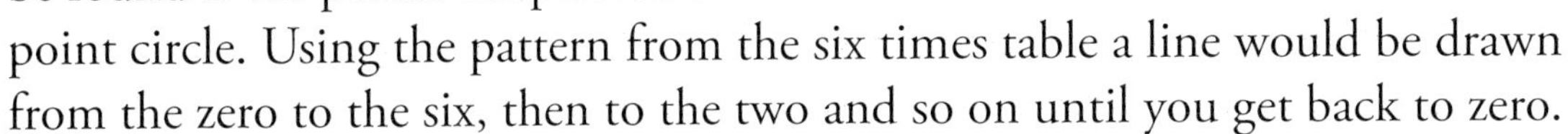

point circle. Using the pattern from the six times table a line would be drawn from the zero to the six, then to the two and so on until you get back to zero.

Now try plotting the four times table on the circle below.

What do you notice?

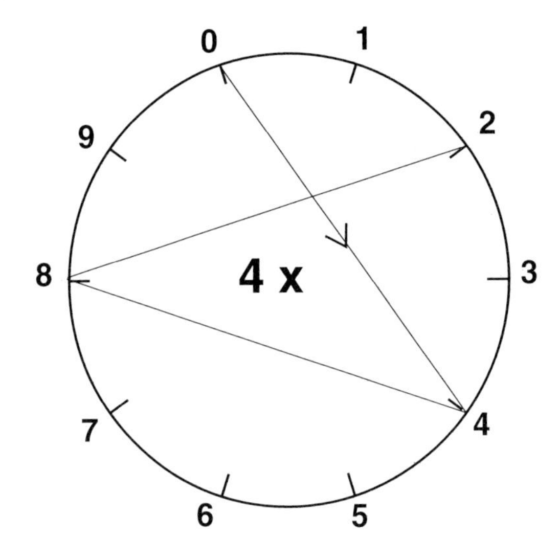

Circular Multiple Mazes

Try plotting the rest of the tables onto ten point circles.

What do you notice about the patterns?

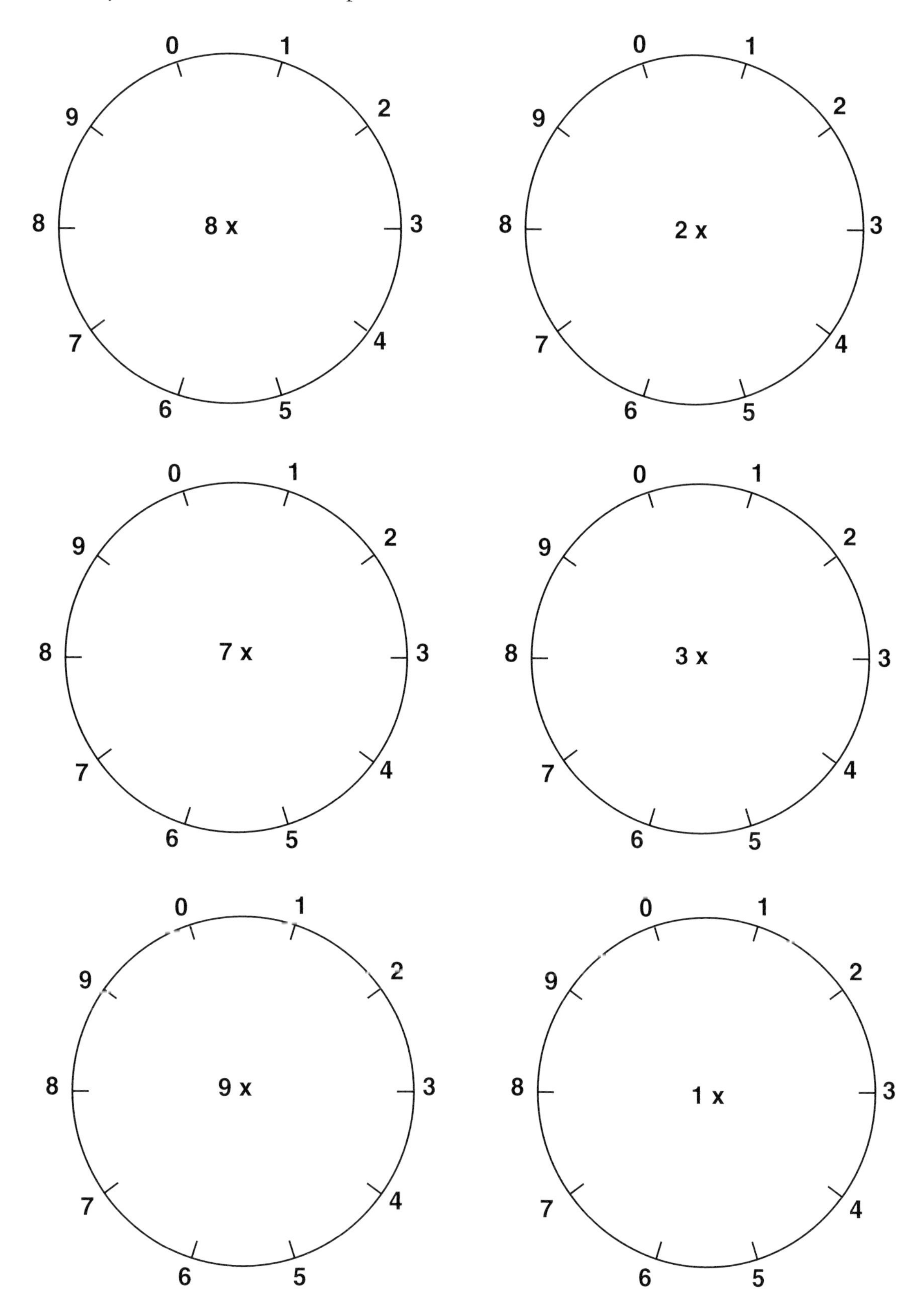

All of The Answers

All of the answers to the table facts from 1 x 1 to 10 x 10 have been placed in ascending order (that is, from smallest to largest) on the chart below.

Complete the chart by filling in the missing numbers.

Some have already been done for you. For example, 1 x 1 = 1, 1 x 2 = 2, 1 x 3 = 3, 1 x 10 = 10 and 2 x 5 and 5 x 2. Fifteen is made by multiplying 3 by 5 (and 5 x 3) and so on.

1	2	3			
			10		
15					24
					35
			60		
					100

Explain why there are more facts than answers.

Product Pairs

The answers to all the tables from 1 x 9 to 9 x 9 are shown in the table below. Write a table fact to match the answer in the same rectangle as the answer. For example, 2 x 5 may be matched to the answer 10.

1	2	3	4
5	6	7	8
9	10 2 x 5	12	14
15	16	18	20
21	24	25	27
28	30	32	35
36	40	42	45
48	49	54	56
63	64	72	81

Marking Multiples

Colour in all the multiples of 4 on the number grid. Some have already been marked in.

What patterns do you notice? Describe them.

Now try colouring in the multiples of eight on the second number grid.

What patterns do you notice? Describe them.

Look at the first number grid that shows the multiples of four coloured in and the second number grid that shows the multiples of eight coloured in. Explain how the four times table and the eight times table are related.

Explain how the two times table is related to the four times table.

Explain how the two times table is related to the eight times table.

Colour in the multiples of three and six on the number grids below and look for links between the two.

1	2	3	4	5	6	7	8	9	10
11	12	13	14	15	16	17	18	19	20
21	22	23	24	25	26	27	28	29	30
31	32	33	34	35	36	37	38	39	40
41	42	43	44	45	46	47	48	49	50
51	52	53	54	55	56	57	58	59	60
61	62	63	64	65	66	67	68	69	70
71	72	73	74	75	76	77	78	79	80
81	82	83	84	85	86	87	88	89	90
91	92	93	94	95	96	97	98	99	100

1	2	3	4	5	6	7	8	9	10
11	12	13	14	15	16	17	18	19	20
21	22	23	24	25	26	27	28	29	30
31	32	33	34	35	36	37	38	39	40
41	42	43	44	45	46	47	48	49	50
51	52	53	54	55	56	57	58	59	60
61	62	63	64	65	66	67	68	69	70
71	72	73	74	75	76	77	78	79	80
81	82	83	84	85	86	87	88	89	90
91	92	93	94	95	96	97	98	99	100

1	2	3	4	5	6	7	8	9	10
11	12	13	14	15	16	17	18	19	20
21	22	23	24	25	26	27	28	29	30
31	32	33	34	35	36	37	38	39	40
41	42	43	44	45	46	47	48	49	50
51	52	53	54	55	56	57	58	59	60
61	62	63	64	65	66	67	68	69	70
71	72	73	74	75	76	77	78	79	80
81	82	83	84	85	86	87	88	89	90
91	92	93	94	95	96	97	98	99	100

1	2	3	4	5	6	7	8	9	10
11	12	13	14	15	16	17	18	19	20
21	22	23	24	25	26	27	28	29	30
31	32	33	34	35	36	37	38	39	40
41	42	43	44	45	46	47	48	49	50
51	52	53	54	55	56	57	58	59	60
61	62	63	64	65	66	67	68	69	70
71	72	73	74	75	76	77	78	79	80
81	82	83	84	85	86	87	88	89	90
91	92	93	94	95	96	97	98	99	100

Marking Multiples

Colour in all the multiples of 9 on the number grid. Some have already been marked in.

What patterns do you notice? Describe them.

Why do the multiples of nine form a diagonal from the top right of the grid to the bottom left of the grid?

Compare the grids that you coloured in showing the multiples of three and the multiples of six with the grid showing the multiples of nine. What do you notice?

Predict what you think might happen if you change the width of the grid so that it is eight columns wide and then colour in the multiples of nine.

Experiment with different size grids and different multiples. For example, what happens when you mark the multiples of seven on a grid that is eight columns wide and then mark them on a grid that is 6 columns wide?

1	2	3	4	5	6	7	8	9	10
11	12	13	14	15	16	17	18	19	20
21	22	23	24	25	26	27	28	29	30
31	32	33	34	35	36	37	38	39	40
41	42	43	44	45	46	47	48	49	50
51	52	53	54	55	56	57	58	59	60
61	62	63	64	65	66	67	68	69	70
71	72	73	74	75	76	77	78	79	80
81	82	83	84	85	86	87	88	89	90
91	92	93	94	95	96	97	98	99	100

1	2	3	4	5	6	7	8
9	10	11	12	13	14	15	16
17	18	19	20	21	22	23	24
25	26	27	28	29	30	31	32
33	34	35	36	37	38	39	40
41	42	43	44	45	46	47	48
49	50	51	52	53	54	55	56
57	58	59	60	61	62	63	64
65	66	67	68	69	70	71	72
73	74	75	76	77	78	79	80

1	2	3	4	5	6	7	8
9	10	11	12	13	14	15	16
17	18	19	20	21	22	23	24
25	26	27	28	29	30	31	32
33	34	35	36	37	38	39	40
41	42	43	44	45	46	47	48
49	50	51	52	53	54	55	56
57	58	59	60	61	62	63	64
65	66	67	68	69	70	71	72
73	74	75	76	77	78	79	80

1	2	3	4	5	6
7	8	9	10	11	12
13	14	15	16	17	18
19	20	21	22	23	24
25	26	27	28	29	30
31	32	33	34	35	36
37	38	39	40	41	42
43	44	45	46	47	48
49	50	51	52	53	54
55	56	57	58	59	60

Multiplication Jigsaw

Multiplication grids can be cut up to produce a jigsaw. In this case the grid does not include the 'zero times table' or the 'ten times table'. The patterns in the remaining table facts may be used to help re-assemble the grid.

The same activity may be varied in many different ways. Students could simply be asked to re-assemble the pieces on a copy of the grid. Students could be allowed to refer to a small version of the grid and put the pieces together. Students could be asked to assemble the pieces without being allowed to refer to the grid.

After the students have tried assembling a multiplication jigsaw, they can cut up a multiplication grid to produce their own jigsaws for other class members to assemble.

After a jigsaw has been created, it may be made more difficult by blanking out some of the numbers. Once some numbers have been blanked out the pieces may be copied and cut up for other students to re-assemble.

Simpler jigsaws may be created by cutting a multiplication grid into 25 two-by-two squares. Deleting some numbers in each two-by-two square will increase the difficulty.

1	2	3	4	5	6	7	8	9
2	4	6	8	10	12	14	16	18
3	6	9	12	15	18	21	24	27
4	8	12	16	20	24	28	32	36
5	10	15	20	25	30	35	40	45
6	12	18	24	30	36	42	48	54
7	14	21	28	35	42	49	56	63
8	16	24	32	40	48	56	64	72
9	18	27	36	45	54	63	72	81

Multiplication Jigsaw

Cut the following multiplication grid along the heavy lines.

Try re-forming the multiplication grid. The patterns in the multiplication grid should help.

1	2	3	4	5	6	7	8	9
2	4	6	8	10	12	14	16	18
3	6	9	12	15	18	21	24	27
4	8	12	16	20	24	28	32	36
5	10	15	20	25	30	35	40	45
6	12	18	24	30	36	42	48	54
7	14	21	28	35	42	49	56	63
8	16	24	32	40	48	56	64	72
9	18	27	36	45	54	63	72	81

Multiplication Jigsaw

Cut the following multiplication grid along the heavy lines.

Try re-forming the multiplication grid. The patterns in the multiplication grid should help.

1		3		5	6			9
		6		10		14		
3	6			15		21		27
	8	12						
5						35	40	
6			24			42		
	14			35		49		63
8	16	24	32				64	
			36		54	63		

Developing Speed and Efficiency

The issue of drill has deliberately been left to this point in the book to emphasise the fact that ***premature drill of basic multiplication facts does more harm than good. Drill does not teach, it simply improves the speed of recall of already known facts.***

Ultimately, however, there comes a time when students need to become fluent with the basic multiplication facts. There is no hiding the fact that this requires effort and commitment on the part of the learner. Remember the only place that success comes before work is in the dictionary!

The National Council of Teachers of Mathematics [in the United States] *Principles and Standards for School Mathematics* comments on this issue.

> Developing fluency requires a balance and connection between conceptual understanding and computational proficiency. (NCTM 2000, p. 35)

There is no doubt that students who have a good grasp of all the basic facts, not just the multiplication facts, develop confidence in their 'mathematical ability'. This is reinforced by family members who equate quick, accurate responses to basic multiplication facts with good ability at mathematics. This in turn fosters a positive attitude in students. ***Likewise if drill activities are introduced prematurely students can learn to hate mathematics.***

The key issues associated with the use of drill are discussed below.

Timed Tests

Of particular concern is the use of timed tests as a means of scaring students into learning their tables at home. Marilyn Burns, a well recognised mathematics educator in the United States, made this comment about the use of timed tests, which summarises the key issues associated with their use.

> Teachers who use timed tests believe that the tests help children learn basic facts. This makes no instructional sense. Children who perform well under time pressure display their skills. Children who have difficulty with skills, or who work more slowly, run the risk of reinforcing wrong learning under pressure. In addition, children can become fearful and negative toward their math learning. (Burns, 2000, p. 157)

Burns, M. (2000). *About teaching mathematics; A K–8 resource* (2nd ed.). Sausalito, CA: Math Solutions Publications.

The following section helps illustrate the point that Burns makes about "reinforcing wrong learning under pressure".

Practice Makes Perfect

There is an old saying that practice makes perfect. This is rubbish. ***Perfect practice is what makes perfect.*** My eldest son became very fast at saying 6 x 9 = 64. No that is not a misprint! He was very good at saying it and very fast – unfortunately he was also wrong. He had been chanting 6 x 9 = 64, while, hopefully, the rest of his class had been chanting 6 x 9 = 54. As a result he had become very quick at stating the product of 6 and 9. It took a great deal of time and effort to rectify this problem. Thankfully, in this case, no connection had been made to 9 x 6. The purpose of the story is to illustrate the danger in rushing into drill and practice activities before the basic fact is clearly established and connections have been made.

Speed versus Accuracy

It is no good being fast at responding to basic multiplication questions if every answer is wrong. Likewise it would be extremely inefficient if it took minutes to give a correct answer to a basic facts question. ***Clearly there is a balance between speed and accuracy that needs to be achieved.***

The emphasis on instant or automatic recall of the basic multiplication facts often means that tables drill becomes a testing situation. The increase in mathematics anxiety that results from this practice often works against developing fluency.

Keith Devlin, author of *The mathematical gene*, noted:

> Despite many hours of practice, most people encounter great difficulty with the multiplication tables. Ordinary adults of average intelligence make mistakes roughly 10% of the time. Some multiplications, such as 8 x 7 or 9 x 7, can take up to 2 seconds and the error rate goes up to 25 percent. (p. 60)

Devlin, K. (2000). *The maths gene: Why everyone has it, but most people don't use it.* London: Weidenfeld & Nicholson.

What About Chanting the Tables?

A favourite cartoon depicts a child singing 1 x 7 is 7, 2 x 7 is 14 and 3 x 7 is 21 and then stopping at 4 x 7 and stating that he had forgotten the tune. While it might be argued from a multiple intelligences perspective that some students are musical/rhythmical learners, children who learn the basic multiplication facts using this approach often have to recite the whole of a set of tables until they reach an individual fact. This is not only inefficient but reveals a lack of understanding on the part of the student and hinders fluency.

Consider also the story of my eldest son who learned to chant 6 x 9 is 64 and became very quick at saying 6 x 9 is 64. Too bad it was wrong! It is often difficult to hear what individual children are saying when the whole class is chanting the tables.

When is it Appropriate to Start Drill & Practice Activities?

Students need to have a good grasp of any basic facts that you plan to drill, before commencing the drill. As the student reaches the transition from derived facts to known facts, he/she is ready to commence drilling those known facts so that memory is strengthened and the speed of recall is increased to the point where the student retrieves the fact without really having to think about the process. The student is now fluent with the particular fact.

What About 'Speed Maths' Methods?

The usefulness of speed maths methods is dependent on whether the students understand why such methods work. When students build personal mental calculation strategies they often make efficient and effective use of them. If they are taught techniques such as 'remove the zeros and add the zeros' without understanding, they can become horribly confused and, while they might learn to calculate quickly, the result may be wrong. Worse still the students may not recognise when the answer is wrong or may generalise to inappropriate contexts. Students may be alerted to alternative calculation strategies but if they adopt a particular method without understanding why it works they will experience difficulty later.

What About Tables Chart?

There tends to be two schools of thought about the use of tables charts. The first suggests that children will see no need to learn the 'table facts' when they have easy reference to a chart of 'table facts'. The second suggestion is that by continually referring to the charts children will begin memorising the individual 'table facts'. It is possible that the frustration of constantly looking for a 'table fact' may encourage children to memorise it. The standard table chart that is laid out in separate columns, however, does little to help children notice patterns and relationships between the 'table facts'.

The Importance of Developing Fluency

Reaching computational fluency with the basic number facts, in this particular case multiplication, means that the cognitive load is decreased in other situations that draw on the use of basic multiplication facts. For example, when estimating the product of 72 and 83 a student will probably round the 72 to 70 and the 83 to 80 and then relate the extended basic fact, 70 x 80 to the fact 7 x 8. Fluency with the basic fact 7 x 8 will free the student to think about the relationship between 7 x 8 and 70 x 80.

Support Activities and Materials

The following activities and games are designed to increase the speed of response to basic multiplication facts, thereby helping to develop fluency. While students engage in these activities it provides opportunity to observe and listen to individual students. Students experiencing problems may then be given assistance.

A set of basic multiplication facts have been provided as a photocopiable resource. You may wish to enlarge this page to A3 size. These multiplication facts are used in several of the games suggested in this section. They are the same facts, with the exception of two 'zero times' facts used in the activity "*I know it, I can work it out, I have no idea*" found on page 26. This activity may be used as an assessment item prior to completing activities in this book and after playing the games in this section. That way you may check on a student's progress.

Note: The facts are written in horizontal form. Some mathematics educators recommend they be written in vertical form to assist in the transition from mental to written methods of calculation. Likewise the photocopiable spinners could also be

Develop Speed and Efficiency

Multiple Turn Over

A game for two players.

You will need two sets of nine playing cards, A(1), 2, 3, 4, 5, 6, 7, 8 and 9.

The cards dealt out in a 3 x 3 array. The second set of nine cards is shuffled and one card is turned over. The player then states the table fact and turns the appropriate card over in the 3 x 3 array.

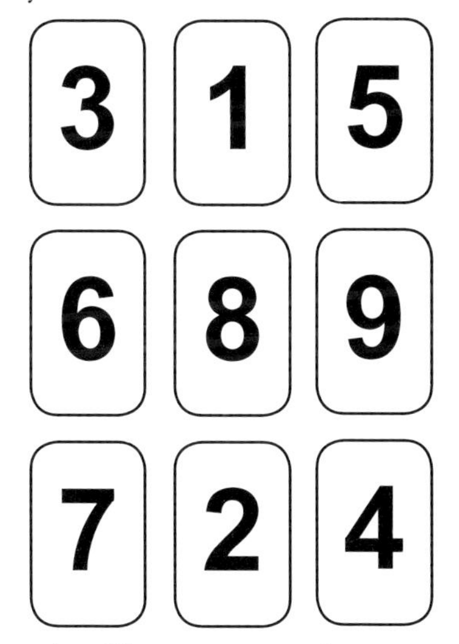

For example, if a seven card is turned over the player would then complete the 'seven times table' in any order. As the table fact is completed the related card is turned over. Play continues until all the cards have been turned over. The cards in the 3 x 3 array are then turned face-up and the processes repeated for the next multiple that is turned over from the second deck of nine cards.

Variations

Deal the cards in the 3 x 3 array face-down so that the student does not have a choice as to the order in which the numbers are to be multiplied.

Rather than focus on all the table facts from 1 to 9, some cards may be removed from the second deck. For example instead of having a second deck containing the cards 1 - 9, you could have the cards 6, 7, 8 and 9 in a second deck containing just four cards.

Flipping Facts

A game for two players

You will need a deck of playing cards with the picture cards (Jack, Queen, King and Joker) removed. Aces represent one.

Roughly halve the deck of cards. Each player is given half the deck. Players simultaneously turn over a card. The first to state the multiplication fact is the winner. For example, if the players turn over 3 and 7, the first player to answer 21 is the winner. The two cards are then kept face down on the table next to the winner. Should there be a tie, the cards are left in the middle and jackpot to the next turn, when the fastest player would win all four cards.

At the end of the game players count their cards to determine who has won the most cards and hence who is the overall winner.

Note: This games needs to be carefully monitored so that students are evenly matched on ability, otherwise one students might continually lose.

Multiplication Match

A game for two - four players

You will need a set of cards containing the table facts that you wish the students to practise. [See BLM, p. 84 for a set of tables in card format.]

Each player is dealt three cards, face down which are left on the table. The first player turns any two of the cards over, stating the product. The player must then guess whether the product on the third (unseen) card will be less than, greater than or in between the two products that have already been stated.

If the player is correct he receives a point. The used cards are not returned to the deck. Three new cards are dealt to the player. Play continues until the deck is exhausted. Shuffle and repeat for a new game.

Battling Basics

A game for 2 - 4 players.

You will need a set of multiplication fact cards. [BLM p. 84.]

The cards are shuffled and the entire deck is dealt out to the players. The cards are left face down in front of each player. Each player then turns over any card and the player with the largest (or smallest) product wins. In order to take the other two cards, however, the player must state the correct answer.

The winning player then places the cards in a separate pile. At the end of the game players count the number of cards they have won in order to determine the winner. If two cards turn up that have the same result, for example, 3 x 8 and 6 x 4, then the cards are left on the table and jackpot for the next turn, when the winner will receive twice as many cards.

Disputes may be settled by checking on a multiplication grid.

Beat the Calculator

A game for 2 - 3 players

You will need a calculator and a set of multiplication cards. [BLM p. 84.]

Basically this is a race against the calculator game. One player has a calculator, while the other works the answer out mentally. The third player is the dealer who turns over the question card and adjudicates on the winner.

The player using the calculator must enter the entire calculation into the calculator and show the answer to the adjudicator. The other player mentally calculates the answer and simply states it to the adjudicator who checks that it is correct. A multiplication grid may be used to verify answers.

The player using mental methods should be able to beat the calculator each time.

The Hidden Card

A game for two players.

You will need a deck of cards with the picture cards removed. Ace represents one. Alternatively you may use two six-sided dice for simple tables and two ten-sided dice for practising harder tables.

Each player is dealt two cards face down. Each player carefully looks at the two cards and multiplies the numbers shown.

PlayerOne chooses one card to turn face up and states the product of the two cards. The second player then tries to work out the number on the hidden card. This would involve working out *factors* of the product that was stated or *using division* (the inverse operation) to determine the value of the hidden card.

If the answer is correct the player wins the two cards and keeps them in a pile. If the player cannot answer or the answer is wrong, these cards are kept in a separate pile as they represent a multiplication/division fact that is still causing a few problems. At the end of the game players can note any facts that are causing them problems and work on them.

If playing with dice, the player rolls the two dice and covers one before the other player sees it.

Show me

A game for small groups or the whole class.

Each player will need two sets of digit cards numbered 2 - 9. Alternatively you may use playing cards. Note: If playing as a whole class the cards need to be large enough to be seen by the teacher some distance away.

The teacher or group leader calls out a number and the players hold up the *factors* that multiply to make the number. Use the numbers from the "*Product Pairs*" activity (p. 71) as the calling list and mark them off as you go. The students will notice that the only time they need to use two of the same numbered card is when a square number is called out.

Count Down

A game for the whole class.

Each students will require a set of numbered cards 2 - 9 large enough to be seen by the teacher at a distance.

The teacher calls out a question similar to ones below. The teacher then begins the count down. Initially the teacher might count down from five, but once the students have developed fluency reduce the count to "3, 2, 1'. Before the count is complete the students must hold up the appropriate number card.

For example, the question might be:

- How many threes in twenty-one? (Students hold up card showing 7.)
- What is twenty-one divided by three? (7)
- Hold up a number that is a factor of twenty-one. (Students might hold up 7 or 3)
- Hold up two numbers that are factors of twenty-one. (7 and 3)
- Hold up a number that divides exactly into twenty-one. (7 or 3)
- What number do I multiply 3 by to get 21? (7)
- What number do I divide 21 by to get 7? (3)
- What is twenty-one divided by three? (7)
- What is $^1/_3$ of twenty-one? (7)

Substitute appropriate numbers into these questions to develop you own question for different families of facts.

Notice how these questions emphasis fact families and the link between multiplication and division. See "*Making the Links*" (p. 38).

If some students feel pressured by the count down aspect of this game, change the name to "Hold Up" and subtly increase the speed of response required.

Doubling and Halving

A game for two – four players.

You will need a deck of cards with the picture cards removed. The ace represents one.

The deck is placed face down in the middle of the table within easy reach of all the players.

The first player picks up a card from the deck.

If the card shows an odd number the number is doubled.

If the card shows and even number it is halved.

The card is replaced at the bottom of the deck and the second player then draws a card

A running total is kept. The first player to reach 100 is the winner. (The total may be adjusted.)

Variations

Create you own deck of cards using the larger numbers. Adjust the total accordingly.

Square Scare

A game for two – four players.

You will need a ten-sided dice (or a set of playing cards ace – 9) and pencil and paper to keep a record.

The first player rolls the dice (or draws a card from a container) and squares the number that is shown.

Players take turns repeating the process.

Players keep a running total. The first player to pass 500 is the winner.

Sand timers may be used to encourage students to increase their speed of response, that is 'beat the timer'. Sand timers that measure different times may be purchased.

Spin A Fact

A game for two players.

You will need a copy of the 'Spin a Fact' spinners. [See p. 85]

There are several ways the spinners might be used. For example, when using the multiplication spinner, 0 - 9, 0 - 9, one side of the spinner can be fixed on a particular number so that specific multiplication facts may be learned. Fixing the left spinner on nine and spinning the right spinner will provide practice in multiplying nine by the numbers 0 - 9. If a students needs to work on their 'seven times table', then the left spinner may be 'fixed on seven'.

To highlight the commutative property of multiplication, the right spinner may be fixed and the left spinner, spun to produce random numbers to multiply.

Variations

Once students are secure in their knowledge of the basic multiplication facts it makes sense to extend these facts to multiply numbers like 70 and 4, where this multiplication may be related to the appropriate basic multiplication fact, 7 x 4. Spinners that focus on these extended basic facts have been included. In order to make the transition from the basic multiplication fact to the extended basic fact students will need to have developed their understanding of place value.

Making the spinners

Make the spinners by punching a hole through the centre circles of each spinner and inserting a plastic spinner arrow.

If plastic spinner arrows are not available you may use a paper clip and a pen or pencil to make a spinner arrow. Place the paper clip on the spinner. Put the point of the pencil through the end of the paper clip and place the point on the centre of the spinner. Flick the paper clip. It will spin around the point of the pen or pencil.

Roll a Fact

A game for two players, each with a different coloured pen.

You will need two dice (0 – 9), two different coloured pens and a blank multiplication fact chart. [You can use the grid from "Making a Tables Map", p. 21. Note: Students will not use the column and row reserved for the 'ten times table'.]

Player one rolls the two dice. The numbers shown on the two dice indicate the factors. If a double comes up the player may only record the product in one place on the grid. Other facts may be recorded in one of two places. For example, if a 5 and 4 turn up, the player may choose to multiply 5 by 4 and enter 20 in the appropriate spot, or multiply 4 by 5 and enter twenty in the appropriate place.

Play continues until one player completes three facts in a row, either vertically, horizontally or diagonally. The game may be made longer by changing it to four facts in a row.

Multiply and Add

A game for two - four players.

You will need one ten-sided dice (0 – 9).

Each round a new multiplier is chosen from the numbers 2 – 9. The game continues until each multiplier has been used once. If 4 is chosen as the multiplier then it cannot be used any more for that game.

The first player rolls the dice and multiplies the number shown by the multiplier (in this case 4) and adds the number 4 to the total. For example, if player 1 rolls 6, then the score for that player would be 6 x 4 + 4 = 28 (or the same as 7 x 4). Students may make the connection that a quick way to calculate the total is to multiply by one more than the number shown on the dice.

Play continues until all the numbers 2 – 9 have been exhausted. Each player keeps a running total of their score for the round. The player with the highest total is the winner.

Answers

Last Digit Patterns (p. 66)

Many patterns exist within the table. The second (the row that begins with 1) and the ninth rows are the reverse of each other. A similar pattern exists in the third and eighth rows. Patterns exist in the columns as well. The digit patterns for the 13, 23 and 33 times table are the same as the three times table. The four times table is the same as the 14, 24, the 34 times table and so on.

x	0	1	2	3	4	5	6	7	8	9	10
0	0	0	0	0	0	0	0	0	0	0	0
1	0	1	2	3	4	5	6	7	8	9	0
2	0	2	4	6	8	0	2	4	6	8	0
3	0	3	6	9	2	5	8	1	4	7	0
4	0	4	8	2	6	0	4	8	2	8	0
5	0	5	0	5	0	5	0	5	0	5	0
6	0	6	2	8	4	0	6	2	8	4	0
7	0	7	4	1	8	5	2	9	6	3	0
8	0	8	6	4	2	0	8	6	4	2	0
9	0	9	8	7	6	5	4	3	2	1	0
10	0	0	0	0	0	0	0	0	0	0	0

The Ones Digit (p. 67)

The most likely digit(s) to appear in the ones place are 2, 6 and 8.

x	1	2	3	4	5	6	7	8	9
1	1	2	3	4	5	6	7	8	9
2	2	4	6	8	0	2	4	6	8
3	3	6	9	2	5	8	1	4	7
4	4	8	2	6	0	4	8	2	8
5	5	0	5	0	5	0	5	0	5
6	6	2	8	4	0	6	2	8	4
7	7	4	1	8	5	2	9	6	3
8	8	6	4	2	0	8	6	4	2
9	9	8	7	6	5	4	3	2	1

Chance result is odd 25/81.□ Chance result is even 56/81.

All The Answers (p. 70)

1	2	3	4	5	6
7	8	9	10	12	14
15	16	18	20	21	24
25	27	28	30	32	35
36	40	42	45	48	49
50	54	56	60	63	64
70	72	80	81	90	100

Product Pairs (p. 71)

- 1 x 1 = 1,
- 1 x 2 and 2 x 1 = 2
- 1 x 3 and 3 x 1 = 3
- 1 x 4, 4 x 1 and 2 x 2 = 4
- 1 x 5 and 5 x 1 = 5
- 1 x 6 , 6 x 1, 2 x 3 and 3 x 2 = 6
- 1 x 7 and 7 x 1 = 7
- 1 x 8, 8 x 1, 2 x 4 and 4 x 2 = 8
- 1 x 9 , 9 x 1 and 3 x 3 =9
- 1 x 10, 10 x 1, 2 x 5, 5 x 2 =10
- 2 x 6, 6 x 2, 3 x 4, 4 x 3 =12
- 2 x 7 and 7 x 2 = 14
- 3 x 5 and 5 x 3 = 15
- 2 x 8, 8 x 2 and 4 x 4 = 16
- 2 x 9, 9 x 2, 3 x 6, 6 x 3 = 18
- 2 x 10, 10 x 2, 4 x 5, 5 x 4 = 20
- 3 x 7 and 7 x 3 =21
- 3 x 8, 8 x 3, 4 x 6 and 6 x 4 = 24
- 5 x 5 = 25
- 3 x 9 and 9 x 3 = 27
- 4 x 7 and 7 x 4 = 28
- 5 x 6, 6 x 5, 3 x 10, 10 x 3 = 30
- 4 x 8 and 8 x 4 = 32
- 5 x 7 and 7 x 5 = 35
- 4 x 9 , 9 x 4 and 6 x 6 = 36
- 4 x 10, 10 x 4, 5 x 8, 8 x 5 = 40
- 6 x 7 and 7 x 6 = 42
- 5 x 9 and 9 x 5 = 45
- 6 x 8 and 8 x 6 = 48
- 7 x 7 = 49
- 5 x 10 and 10 x 5 = 50
- 6 x 9 and 9 x 6 = 54
- 7 x 8 and 8 x 7 = 56
- 6 x 10 and 10 x 6 = 60
- 7 x 9 and 9 x 7 = 63
- 8 x 8 = 64
- 7 x 10 and 10 x 7 = 70
- 8 x 9 and 9 x 8 = 72,
- 8 x 10 and 10 x 8 = 80
- 9 x 9 = 81
- 9 x 10 and 10 x 9 = 90
- 10 x 10 = 100.

Marking Multiples (p. 72 - 73)

This activity helps to focus children on various patterns within the table facts. For example, the relationship between the □two, four and eight times tables□ and the □three, six and nine times tables□ The geometric pattern formed by the multiples of nine illustrate the relationship between the nine and ten times table. i.e. 1 x 9 = 1 x 10 □ 1, 2 x 9 = 2 x 10 □ 2, 3 x 9 = 3 x 10 □ 3. . . In addition, when shading in the charts, children use a variety of strategies such as skip counting.

When the nine-times table is shaded on the 1 □ 80 chart the diagonal will run from the top left to the bottom right. This occurs because the nine-times table is related to the eight -times table. i.e. 1 x 9 = 1 x 8 + 1, 2 x 9 = 2 x 8 + 2, 3 x 9 = 3 x 8 + 3.

When the □seven times table□ is shaded on the eight column grid it will form a similar diagonal pattern (top right to bottom left) to the multiples of nine on the ten column grid. When the multiples of seven are marked on a six column grid a diagonal pattern from the top left to the bottom right of the grid is formed (similar to the multiples of nine, when they were marked on an eight column grid).

Multiplication Match Cards

3 x 3	7 x 2	4 x 3	3 x 5	8 x 7	4 x 7
6 x 2	3 x 4	5 x 2	4 x 4	2 x 8	4 x 2
7 x 3	8 x 4	2 x 9	6 x 7	2 x 3	9 x 2
9 x 4	6 x 6	3 x 8	7 x 4	7 x 5	9 x 8
0 x 8	5 x 8	5 x 5	8 x 6	9 x 3	5 x 4
5 x 6	9 x 9	7 x 8	8 x 3	3 x 9	2 x 7
8 x 8	4 x 6	9 x 6	4 x 9	2 x 6	4 x 8
6 x 9	8 x 5	2 x 5	7 x 9	5 x 9	8 x 9
1 x 7	3 x 7	2 x 4	8 x 2	9 x 5	6 x 8
7 x 7	2 x 2	7 x 6	6 x 5	6 x 4	5 x 3
5 x 7	4 x 5	3 x 2	3 x 6	6 x 3	9 x 7

Spin A Fact

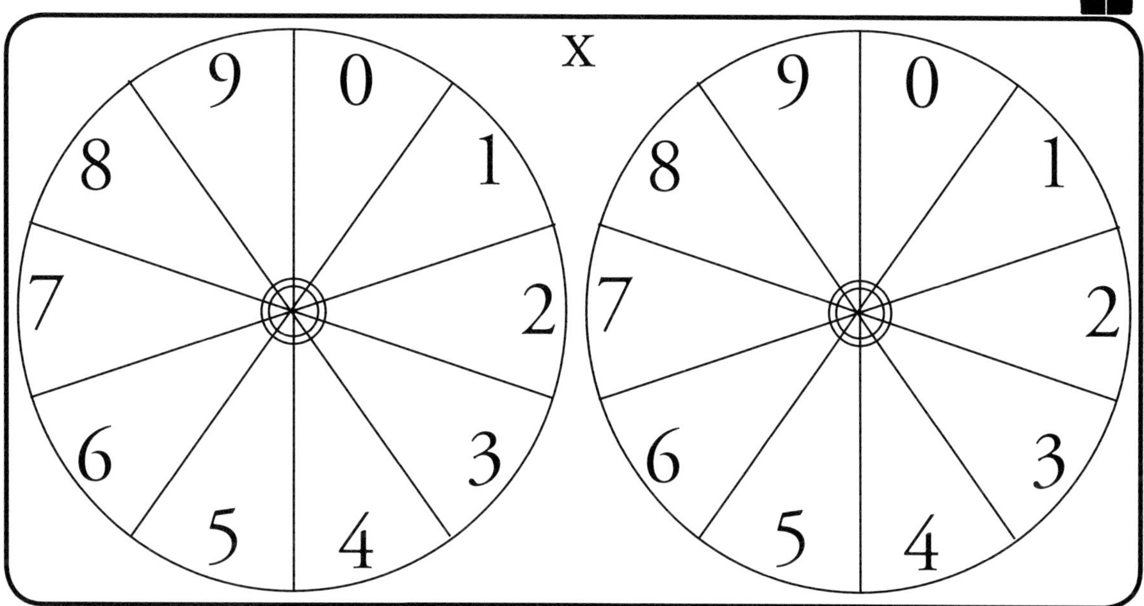
x
9
0
1
2
3
4
5
6
7
8
9
0
1
2
3
4
5
6
7
8

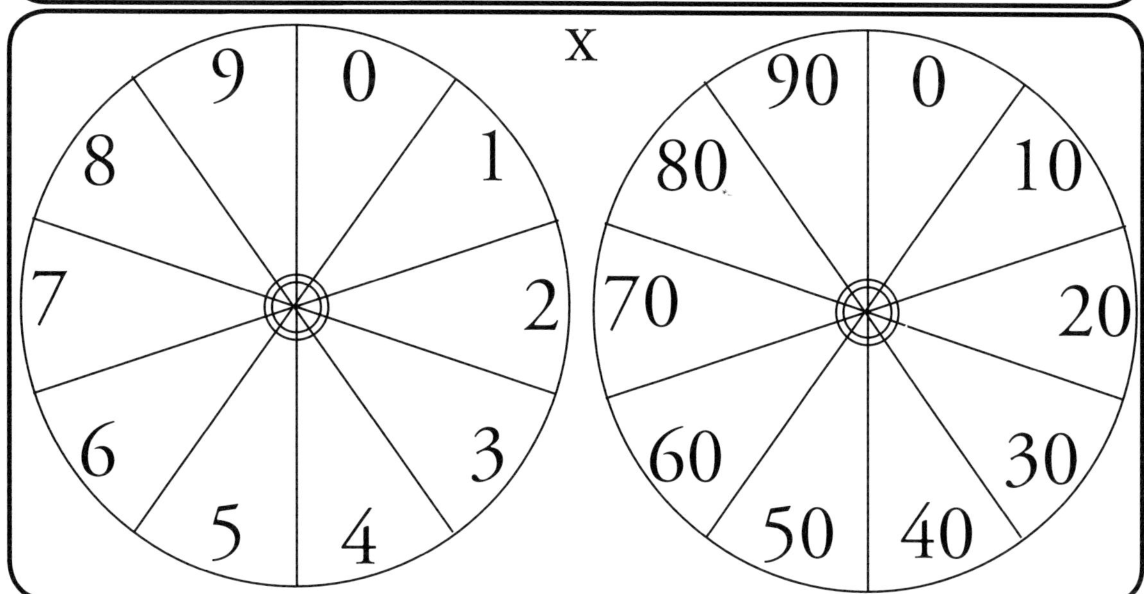
x
9
0
1
2
3
4
5
6
7
8
90
0
10
20
30
40
50
60
70
80

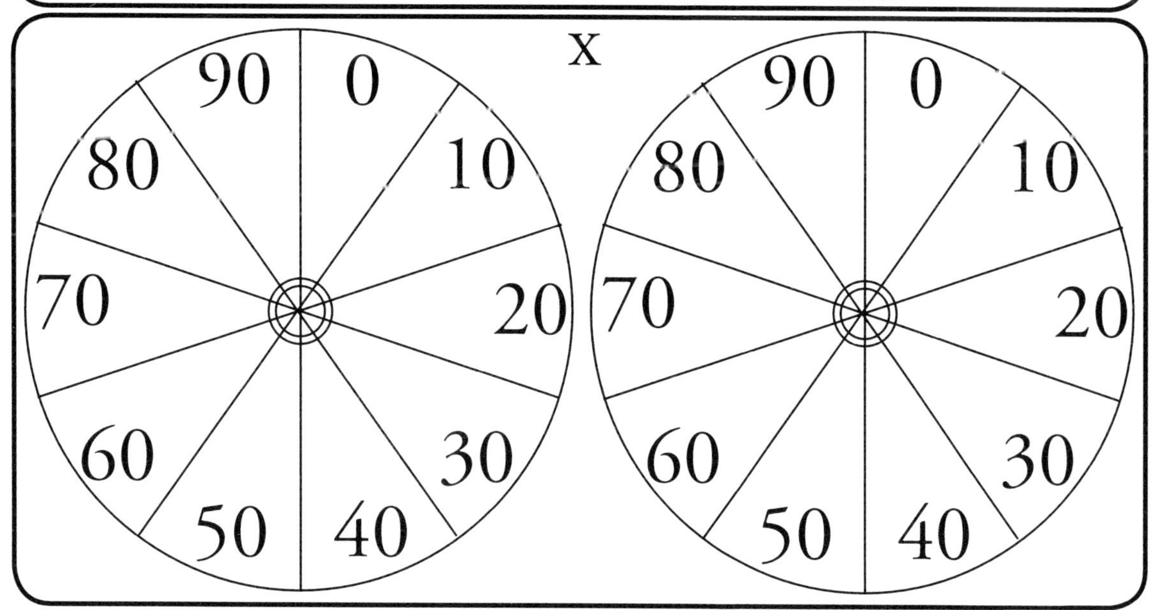
x
90
0
10
20
30
40
50
60
70
80
90
0
10
20
30
40
50
60
70
80

Multiples Grid

1	2	3	4	5	6	7	8	9	10
2	4	6	8	10	12	14	16	18	20
3	6	9	12	15	18	21	24	27	30
4	8	12	16	20	24	28	32	36	40
5	10	15	20	25	30	35	40	45	50
6	12	18	24	30	36	42	48	54	60
7	14	21	28	35	42	49	56	63	70
8	16	24	32	40	48	56	64	72	80
9	18	27	36	45	54	63	72	81	90
10	20	30	40	50	60	70	80	90	100

1	2	3	4	5	6	7	8	9	10
2	4	6	8	10	12	14	16	18	20
3	6	9	12	15	18	21	24	27	30
4	8	12	16	20	24	28	32	36	40
5	10	15	20	25	30	35	40	45	50
6	12	18	24	30	36	42	48	54	60
7	14	21	28	35	42	49	56	63	70
8	16	24	32	40	48	56	64	72	80
9	18	27	36	45	54	63	72	81	90
10	20	30	40	50	60	70	80	90	100

1	2	3	4	5	6	7	8	9	10
2	4	6	8	10	12	14	16	18	20
3	6	9	12	15	18	21	24	27	30
4	8	12	16	20	24	28	32	36	40
5	10	15	20	25	30	35	40	45	50
6	12	18	24	30	36	42	48	54	60
7	14	21	28	35	42	49	56	63	70
8	16	24	32	40	48	56	64	72	80
9	18	27	36	45	54	63	72	81	90
10	20	30	40	50	60	70	80	90	100

Small Tables Charts

x	0	1	2	3	4	5	6	7	8	9
0	0	0	0	0	0	0	0	0	0	0
1	0	1	2	3	4	5	6	7	8	9
2	0	2	4	6	8	10	12	14	16	18
3	0	3	6	9	12	15	18	21	24	27
4	0	4	8	12	16	20	24	28	32	36
5	0	5	10	15	20	25	30	35	40	45
6	0	6	12	18	24	30	36	42	48	54
7	0	7	14	21	28	35	42	49	56	63
8	0	8	16	24	32	40	48	56	64	72
9	0	9	18	27	36	45	54	63	72	81

x	0	1	2	3	4	5	6	7	8	9
0	0	0	0	0	0	0	0	0	0	0
1	0	1	2	3	4	5	6	7	8	9
2	0	2	4	6	8	10	12	14	16	18
3	0	3	6	9	12	15	18	21	24	27
4	0	4	8	12	16	20	24	28	32	36
5	0	5	10	15	20	25	30	35	40	45
6	0	6	12	18	24	30	36	42	48	54
7	0	7	14	21	28	35	42	49	56	63
8	0	8	16	24	32	40	48	56	64	72
9	0	9	18	27	36	45	54	63	72	81

x	0	1	2	3	4	5	6	7	8	9
0	0	0	0	0	0	0	0	0	0	0
1	0	1	2	3	4	5	6	7	8	9
2	0	2	4	6	8	10	12	14	16	18
3	0	3	6	9	12	15	18	21	24	27
4	0	4	8	12	16	20	24	28	32	36
5	0	5	10	15	20	25	30	35	40	45
6	0	6	12	18	24	30	36	42	48	54
7	0	7	14	21	28	35	42	49	56	63
8	0	8	16	24	32	40	48	56	64	72
9	0	9	18	27	36	45	54	63	72	81

x	0	1	2	3	4	5	6	7	8	9
0	0	0	0	0	0	0	0	0	0	0
1	0	1	2	3	4	5	6	7	8	9
2	0	2	4	6	8	10	12	14	16	18
3	0	3	6	9	12	15	18	21	24	27
4	0	4	8	12	16	20	24	28	32	36
5	0	5	10	15	20	25	30	35	40	45
6	0	6	12	18	24	30	36	42	48	54
7	0	7	14	21	28	35	42	49	56	63
8	0	8	16	24	32	40	48	56	64	72
9	0	9	18	27	36	45	54	63	72	81

x	0	1	2	3	4	5	6	7	8	9
0	0	0	0	0	0	0	0	0	0	0
1	0	1	2	3	4	5	6	7	8	9
2	0	2	4	6	8	10	12	14	16	18
3	0	3	6	9	12	15	18	21	24	27
4	0	4	8	12	16	20	24	28	32	36
5	0	5	10	15	20	25	30	35	40	45
6	0	6	12	18	24	30	36	42	48	54
7	0	7	14	21	28	35	42	49	56	63
8	0	8	16	24	32	40	48	56	64	72
9	0	9	18	27	36	45	54	63	72	81

x	0	1	2	3	4	5	6	7	8	9
0	0	0	0	0	0	0	0	0	0	0
1	0	1	2	3	4	5	6	7	8	9
2	0	2	4	6	8	10	12	14	16	18
3	0	3	6	9	12	15	18	21	24	27
4	0	4	8	12	16	20	24	28	32	36
5	0	5	10	15	20	25	30	35	40	45
6	0	6	12	18	24	30	36	42	48	54
7	0	7	14	21	28	35	42	49	56	63
8	0	8	16	24	32	40	48	56	64	72
9	0	9	18	27	36	45	54	63	72	81

Tables Chart

x	0	1	2	3	4	5	6	7	8	9
0	0	0	0	0	0	0	0	0	0	0
1	0	1	2	3	4	5	6	7	8	9
2	0	2	4	6	8	10	12	14	16	18
3	0	3	6	9	12	15	18	21	24	27
4	0	4	8	12	16	20	24	28	32	36
5	0	5	10	15	20	25	30	35	40	45
6	0	6	12	18	24	30	36	42	48	54
7	0	7	14	21	28	35	42	49	56	63
8	0	8	16	24	32	40	48	56	64	72
9	0	9	18	27	36	45	54	63	72	81